Krupin's
Toll·Free
Environmental
Directory

A comprehensive nationwide 800 phone number listing of over seven thousand environmental firms, organizations, government agencies and private institutions that every career seeker and working professional can call long distance for free.

Paul J. Krupin

Published and distributed by:

Direct Contact Publishing, P. O. Box 6726, Kennewick, Washington 99336.

Additional copies may be ordered directly from Direct Contact Publishing.

First Edition First Printing 1994, Second Printing 1995
Second Edition First Printing December 1995

Cover Design by William Leahy Advertising.

Printed in the United States of America

ISBN 1-885035-02-0

TABLE OF CONTENTS

INTRODUCTION

Congratulations! You have in your hands one of the most useful reference books on the environment available. Krupin's Toll-Free Environmental Directory is a powerful communications tool, designed to provide you with instant access to information, technology, expert environmental consulting and engineering services, government agencies, regulatory authorities and officials, professional and non-profit environmental organizations, employment and business opportunities in every major aspect of the environmental industry.

Krupin's Toll-Free Environmental Directory can place you in contact with companies, organizations, associations and government agencies all across the nation virtually for free.

Learn about environmental problems, challenges, strategies and solutions while you talk with knowledgeable people in the industry. You can save phenomenal time, effort and money by researching at home or in your office, and by spending your time, energy and money more wisely, cost-efficiently and effectively. This directory will serve as a source acquisition tool for public, college and school libraries as they build collections of free and inexpensive environmental information and educational materials.

You can improve your purchasing and contracting alternatives and the selection of products, services and opportunities that may not be available to you in your area by reaching out and contacting manufacturers, technologists, merchandisers, suppliers of services nation-wide.

Career seekers can reach out across the nation and utilize informational and research interviewing techniques to enhance their knowledge of the potential employment and network their way to success. Technologists and working professionals can access other professionals in government and industry to pursue business development opportunities, technology transfer information, and innovative solutions to new and challenging environmental situations.

Save money by calling toll-free and eliminating charges for long distance phone calls.

We Put the World At Your Fingertips

Major companies, organizations, government agencies, that you've heard of along with thousands of other firms, businesses, and institutional organizations are all just a toll-free phone call away.

Just look for the subject, and companies or organizations or sources of information in Krupin's Toll-Free Directory and call ... it's that simple. The most effective way to benefit from the access you have is to ask for a specific person, or type of expert. Then Make a list of questions. Ask to speak with technical experts to address your specific needs or problems. Ask Questions. Ask for the names of other people you can talk to. The appendices contain the job titles and job classifications of many of the experts you may wish to contact. Utilize these contact titles to network your way through the industry.

Call around the country and broaden your knowledge of people, technology, products and services you are familiar as well as hundreds of obscure items, ideas, and capabilities you may not ever have heard of before.

Call around the country to compare prices, techniques, technology, and information.

Call and request company literature, catalogs, product information, informational reports, brochures, pamphlets and other data and information. There is an absolutely phenomenal amount of free information available if only you ask. Many companies and organizations will supply free reports and offer helpful advice or guidance in response to your specific questions.

Many of the companies listed under "Eco-preneurs" sell new and unique products. Purchasing products for personal or business use or sending gifts is easy. Just call the toll-free number and use your credit card and you'll have access to unlimited gift giving opportunities. Most companies can deliver your packages overnight if you request.

Using the Directory is Easy!

Check to see the listing to find out if the toll-free number can be called from your area. Most of the toll-free numbers can be called from anywhere in the United States. Occasionally you may have difficulty in being connected with a toll-free number. Check the number with the AT&T 800 Directory Assistance operators at 1-800-555-1212.

Be aware that many companies man their toll-free numbers only during the normal business hours for the time zone in which they are located. You can often leave your name address and phone number and a short message with the companies answering service, if you cannot call back during normal business hours.

Please note that some of the listings are for wholesalers and not retailers. Some of the listings may also only provide the name of the dealer or suppliers of products or services in your area. Although a listing may be for orders only, you can usually request a catalog or information and prices.

You should also be aware that some of the Hotline listings are emergency numbers that are only to be used, for example, in the event you discover and wish to report an observed poaching incident or violation, wish to report a spill or release of hazardous substances to the environment, or are contacting a Poison Control Center about personal exposure to a hazardous or toxic substance. Your cooperation with government authorities is greatly appreciated.

Although we have exhaustively researched all sources to ensure the accuracy and completeness of the information contained in this book, we are always looking for new listings, ways to correct or improve the entries, and ideas and suggestions on how to make this Directory easier to use and more fun to use.

Krupin's Toll-Free Directories are updated with every printing as new information is made available to us. New listings come out continually and some companies' toll-free numbers change from time to time.

So if you want to be listed, would like us to develop a directory on a special topic of interest or have any comments please write to:

DIRECT CONTACT
P. O. Box 6726
Kennewick, Washington 99337

THE DIRECT CONTACT METHOD to Uncover Hidden Jobs & Business Opportunities - By Phone

by Paul J. Krupin, JD

Target the job or opportunity you want, learn everything you need to about the company, call the person who will hire you directly -- all using the DIRECT CONTACT METHOD. Here's how:

THE DIRECT CONTACT METHOD FOR ACHIEVING SUCCESS IN FINDING YOUR TARGET JOB OR OPPORTUNITY:

1. Utilize specialized sources of information - this includes the Yellow Pages, trade and industry publications and regulatory agency documents, and agency administrative files;

2. Identify the names, addresses and phone numbers of potential employers, decision-makers to contact;

3. Prepare to network by doing research and preparing lists of questions;

4. Make contact and ask specific questions - for guidance, advice and names of who else to call;

5. Follow-up within 24 to 48 hours with: a thank you letter, a tailored resume, and other supporting materials

DON'T ASK FOR A JOB -- ASK FOR INFORMATION *** ASKING FOR INFORMATION WILL LAND YOU JOBS

How to conduct your research and where to find information resources are the two biggest hurdles you face. You must overcome any timidness or reluctance you may have about contacting people you don't know. You must begin to make DIRECT CONTACT and practice making DIRECT CONTACT until you acquire capability, comfort and confidence in the style of your approach. DIRECT CONTACT research is not enigmatic or complex. There are just two basic skills required: 1. the ability to read and 2. the ability to talk with people.

RESEARCH BEFORE YOU APPROACH A HIRING AUTHORITY

The best sources of information will be those who work immediately below the decision-making level. The working staff includes the professionals, assistants, or temporaries who work for a hiring authority. These people can give you valuable information on the informal hiring process. The primary goals for your research are to: Identify the people to contact; Identify the people to meet with; Acquire advice;

Receive referrals to other people; Identify the hiring authority and what his or her problems and are.

Your research is successful if you identify that specific organization and that key person who can offer you a job. "Critical information" is readily recognized when you hear it: "Ms. Smith just got funding for a new project..." or "Joe Jones just got promoted and his boss is desperate for a replacement", or "Ms. White opening a new branch office..." or "ABC Co. just was awarded a contract." It's that simple.

WHO CAN YOU REACH & WHEN: Call 8 AM to 5 PM to get staff members to arrange an appointment with a busy or otherwise unavailable individual hiring authority. Call before 8 AM and after 5 PM and you are likely to hit the hiring authorities. Take note -- the boss usually is the first person in the office and the last one to leave, especially in small companies. Before and after hours the pressure is off and they're usually willing to talk.

HOW TO GET IN THE FRONT DOOR AND PAST THE GUARDIANS OF THE ORGANIZATION? These three simple techniques are a well kept secret! 1. You get past the secretary or receptionist by asking to speak with a technical specialist about a technical question. Try it next time you call! Target the technical specialists who have the information you need to have an intelligent interaction with the hiring authority: a. the people who have the job you want; b. the people who had the job you want; c. the people who work for the hiring authority. If you don't know their names -- you ask for them by title. 2. You get more information, more cooperation, better and willing assistance from the office who takes care of the corporate image -- THE PUBLIC RELATIONS OFFICE. Never rely on personnel or human resources. They are just executors of policy and procedure. 3. Ask to speak with the LIBRARIAN. Paid problem solvers -- they either know the answer or where and how to get it.

INFORMATIONAL INTERVIEWING = RESEARCH INTERVIEWING = NETWORKING

Targeted networking focuses on quickly finding specific people who can offer you a job, and then following up effectively. The key is to research and ask the right questions.

Susan Roane, author of "The Secrets of Savvy Networking" (this is a book worth buying and reading twenty times) says networking is: " A reciprocal process, an exchange of ideas, leads and suggestions that support both professional and personal lives. There is also a spirit of sharing that transcends the information shared. The best networkers reflect that spirit with a genuine joy in their "giving. Network" is now an action verb. Follow-up is the magic that makes the process of networking work. Respect for people is the cornerstone of communication and networking."

In my experience -- **GIVING** is the magic that makes the process of networking work. You know how you feel when someone helps you unasked. You're thankful and you feel obligated to help out in return, as soon as you can. To make the best impression and to create a sense of obligation in your prospect, YOU MUST GIVE HIM OR HER SOMETHING WHICH HELPS THEM SOLVE THEIR PROBLEM. It needn't be big, just a relevant and timely DEMONSTRATION OF WHAT YOU COULD DO IF YOU WERE HIRED, to have the impact you desire. You show them "THIS IS WHAT YOU GET! -- HERE'S WHAT'S IN IT FOR YOU!".

NETWORKING WORKS BECAUSE IT OVERCOMES THE INNATE HESITATION OF ACCEPTANCE

Networking involves getting others to: act as a reference, suggest referrals, contact employers on your behalf, recommend other people who can help you, supply information, and make introductions on your behalf. Thus when you network you are fishing for third party endorsements. AVOID QUESTIONS THAT PATENTLY REVEAL A PURELY SELFISH MOTIVE

Remember -- Don't ask for work, ask for information. Information will get you jobs.

ASK OPEN ENDED QUESTIONS. Ask about the informal hiring process, about the internal working operation and conditions of the organization, and about how the organization really functions. You will in the process find out the names you need as well a host of critical data.

WHEN YOU FOLLOW-UP WITH YOUR LETTERS AND RESUME, YOU THEN TELL THEM:

1. the kinds of challenges you like to deal with;
2. the skills you have to deal with those challenges;
3. why you like their organization
4. what kind of employment you are looking for and why. YOU NEED TO DO THIS WITHIN 24 TO 48 HOURS.

Remember that your tailored personalized letter, tailored resume and supporting materials are themselves marketing documents. All marketing documents must have objectives and must want to make the receiver take an identified action such as: a. ask for more information; b. call for an interview; or c. offer you a job. Thus, your letter, resume and supporting materials have these objectives -- to:

1. establish a bond between you and your recipient, marketer and prospect;

2. get the prospect to respect, admire and benefit from what you send, so that there is a desire for future contacts between you and him or her;

3. create a sense of obligation or thankfulness -- put the recipient in your debt for the benefits you have bestowed by writing;

4. give the prospect something of value that he or she will not throw away.

To achieve these objectives, address the recipient directly, with first person "you" and "I" language; remind the prospect right from the start what he or she wants to achieve and lead him or her through a series of power-packed paragraphs.

Here are some simple ideas for how you can do this: Write a short problem solving article summary based on some aspect you really know well. Describe and use your niche and help the employer for free. Show what you could do (if you were to receive the job). Go to the library -- photocopy technical articles and send them as attachments -- make sure you indicate in your letter that you researched them after hearing about the prospects problem.

THIS IS A COMPLETE REPRINT OF ARTICLE IN Fall 1995, "Damn Good Resume Pro Newsletter", published by Yana Parker, author of "The Damn Good Resume", and "Resume Pro: The Professionals Guide", based on the new book "Get In Direct Contact" by Paul J. Krupin

----

This article is based upon his new book "Get in Direct Contact". $14.95 plus $3.00 shipping and handling. To order call 1-800-457-8746 or write to Direct Contact Publishing, P. O. Box 6726, Kennewick WA 99336.

AIR CLEANING & PURIFICATION

A ir Purification Systems	San Diego CA	1-800-776-6746
Abatement Technologies	Lawrenceville GA	1-800-634-9091
Action Clean Air Services Inc	Orlando FL	1-800-524-3042
Air & Water Concepts	Flora IL	1-800-662-3842
Air Cleaners For Allergy Problems	Houston TX	1-800-321-1096
Air Cleaning & Purification	Conyers GA	1-800-508-1776
Air Cleaning Specialists	Houston TX	1-800-544-5380
Air Cleaning Systems Inc	Latham NY	1-800-247-1020
Air Climate Systems	Minneapolis MN	1-800-292-7733
Air Purification Systems	Fridley MN	1-800-521-7369
Air Quality Engineering	Minneapolis MN	1-800-328-0787
Air Quality Product	Springfield IL	1-800-837-1514
Air Safe Inc.	Grand Haven MI	1-800-723-0030
Air Technology Intl Inc	Wichita KS	1-800-447-9942
Allergy Fighter By Breathe-Eze	Arlington TX	1-800-247-7316
Alpine Air Products	Hastings MN	1-800-437-1960
Alpine Air	Needham Heights MA	1-800-628-2209
Alpine Industries	Hershey NE	1-800-262-9508
Alpine Industries	LaPorte CO	1-800-551-4344
Alpine Sales	Medford NJ	1-800-345-8802
Aston Technology	Conyers GA	1-800-922-0005
B & R Investments	Ogden UT	1-800-694-6525
Beckert & Hiester	Saginaw MI	1-800-332-4031
Best Air Systems	Gibsonia PA	1-800-354-5804
Bio Climatic Inc	Mt Laurel NJ	1-800-962-5594
Bionaire	Allendale NJ	1-800-253-2764
Clean Air Inc	Norwalk CT	1-800-242-9424
Clean Air Systems Of Brooklyn Park MN	Brooklyn Park MN	1-800-423-9728
Clearveil	Denver CO	1-800-531-6662
Crystal Air/Smoketeer Of Cincinnati OH	Cincinnati OH	1-800-551-5401
Dakota Air & Equip	Arlington SD	1-800-247-6235
Dareco/Air Filters	Blaine MN	1-800-533-9461
Dexon Mfg Inc	Rush City MN	1-800-322-7604
Dunnack Enterprises	Andover CT	1-800-742-0631
Dustvent	Addison IL	1-800-553-3687
E L Foust Co	Elmhurst IL	1-800-225-9549
Emerald Environmental Inc	Roslyn Heights NY	1-800-300-3951
Envirco	Hagerstown MD	1-800-645-1610
F-Matic Of America	American Fork UT	1-800-824-9994
Filter Queen Of Beaver County	Bridgewater PA	1-800-441-1084
Fluid Handling Specialist	Baton Rouge LA	1-800-433-5799
Fore N Aft Pure Air Systems	Panama City FL	1-800-247-8212
Four Starr Improvement Svcs	Reading PA	1-800-375-2598
Fuel Tech	Stamford CT	1-800-225-0085
Golden Eagle Limited	Dallas TX	1-800-754-2716
Hepa Air Cleaning Equip	Bridgeport CT	1-800-532-5373
Home Environmental Products Co	Stafford TX	1-800-551-4600
Honeywell Environmental Air Control Inc.	Hagerstown MD	1-800-638-7416
Howe Baker Engineers, Inc.	Tyler TX	1-800-323-2115
I A Q Inc	Monarch Beach CA	1-800-699-4368
Indoor Air Technologies	Pickerington OH	1-800-469-6561
Isolaire Corp	Cockeysville MD	1-800-835-3852

Karde Air Purification Systems	Medford NJ	1-800-245-8802
Koger Air Corporation	Martinsville VA	1-800-368-2096
Kurz Instruments	Monterey CA	1-800-424-7356
Lab Air Products & Services Inc	Bartlett TN	1-800-755-8400
Munster-Dyer	Dyer IN	1-800-362-6967
Munters NE Gas Cleaning Division	Amesbury MA	1-800-446-5696
Natura Brand Products	Mission Viejo CA	1-800-582-8442
Natures Quarters	Haddam CT	1-800-798-6699
Netco International Independent Distrib	Eustis FL	1-800-589-4341
New Era Air	Waverly IA	1-800-352-5711
Nonscents Of Louisiana	Jena LA	1-800-457-2224
Northern Air Corp	Raynham MA	1-800-772-4394
Nuclear Consulting Service Inc	Worthington OH	1-800-992-5192
O'Donnell Distributing	Kingwood TX	1-800-203-2220
Permatron	Franklin Park IL	1-800-882-8012
Polk Air Filter Sales	Lakeland FL	1-800-523-8276
Pollution Research & Development	Claremont NH	1-800-426-2611
Protek Environmental	Roy City TX	1-800-782-0553
Purafil, Inc.	Doraville GA	1-800-222-6367
Pure Air Control Services	Clearwater FL	1-800-422-7873
Pure N Natural Systems	Woburn MA	1-800-237-9199
Quality Air Inc	Baltimore MD	1-800-522-5156
Replacement Parts Supply	Trenton NJ	1-800-873-2496
Rich Air Envirosafe	Indio CA	1-800-695-6525
S & G Services	New Douglas IL	1-800-628-0931
Seaco	Sparks NV	1-800-433-8839
Service Tech	Cleveland OH	1-800-992-9302
Silent Duster Inc	Oxford AL	1-800-453-3878
Smokeever Air Cleaning Systems	Cincinnati OH	1-800-832-7110
T P Enterprises	Mobile AL	1-800-249-0519
Tectronic Products	E Syracuse NY	1-800-227-1375
Thurmond Air Quality Systems	Plano TX	1-800-247-7873
Unico Inc	St Louis MO	1-800-527-0896
United Air Specialists Inc	Cincinnati OH	1-800-992-4422
Val Kleen	Fresno CA	1-800-736-0759
Vitarie Corp	Garfield NJ	1-800-447-4344
Zeocrystal Industries	Crestwood IL	1-800-636-7373
Zontec Inc	Ogdensburg NY	1-800-835-6962

AIR POLLUTION CONTROL & AIR QUALITY MONITORING

Air Purification Systems	Fridley MN	1-800-521-7369
Air Science Consultants Inc.	Bridgeville PA	1-800-759-9282
Air Technology Inc.	Fort Mill SC	1-800-822-8040
Airflow Technical Products Inc.	Netcong NJ	1-800-247-2887
Altech Systems Corp.	Moorpark CA	1-800-992-5832
Ambient Engineering Inc.	Matawan NJ	1-800-732-4155
American Environmental Labs Inc.	Leominster MA	1-800-522-0094
American Norit Co. Inc.	Atlanta GA	1-800-641-9245
Ametek Inc.	Newark DE	1-800-222-6789
Analytical Testing Consultants Inc.	Kannapolis NC	1-800-733-3193
Arizona Instrument Corp	Tempe AZ	1-800-528-7411
Assay Technology Inc.	Palo Alto CA	1-800-833-1258
Auburn International	Danvers MA	1-800-255-5008
Baghouse Services Inc.	Los Alamitos CA	1-800-367-1224

BHA Group Inc.	Kansas City MO	1-800-821-2222
Brown John S Company Inc.	Bloomfield CT	1-800-927-5525
Burney the Burner	Highland CA	1-800-272-1243
CAE Exemplar	Palatine IL	1-800-627-0033
CAE Express	Palatine IL	1-800-223-3977
Candess	Wayland MA	1-800-636-0402
Carbonair Inc.	Maple Grove Mn	1-800-526-4999
Carbtrol Corp.	Westport CT	1-800-242-1150
Carus Chemical	Ottawa IL	1-800-435-6856
CECO Filters Inc.	Conshohocken PA	1-800-220-8021
Clean Air Engineering	Palatine IL	1-800-223-3977
Clean Air Engineering	Palatine IL	1-800-627-0033
Clean Air Products	Statesville NC	1-800-206-5222
CSE Inc.	Roseville MN	1-800-279-7645
Davis Water and Waste Industries Inc.	Tellevast FL	1-800-345-3982
Delta Cooling Towers Inc.	Fairfield NJ	1-800-289-3358
Delta Technical Products Co.	Des Plaines Il	1-800-733-5820
Digicolor Inc.	Columbus OH	1-800-848-6448
Donaldson Co.	Minneapolis MN	1-800-365-1331
Ducon Environmental Systems Inc.	Farmingdale NY	1-800-394-4990
Dustvent Inc.	Addison IL	1-800-553-3687
Eco Sensors Inc.	Sante Fe NM	1-800-472-6626
Edwards Engineering Corp.	Pompton Plains NJ	1-800-526-5201
Environmental Equip Svc	Jeffersonville IN	1-800-866-2072
EPG Companies Inc.	Maple Grove MN	1-800-443-7426
Euroclean	Itasca IL	1-800-545-4372
Foxboro	Foxboro MA	1-800-521-0451
Fuller Co.	Bethlehem PA	1-800-523-9482
Grace Emission Control Products	Baltimore MD	1-800-638-3844
Grace Emission Control Products	DePere WI	1-800-558-2884
Graseby Andersen	Atlanta GA	1-800-241-6898
GT Safety Products	Pawtucket RI	1-800-556-7310
Hydrosil Intl Limited	Arlington Heights IL	1-800-787-7531
Ingold Electrodes Inc.	Wilmington MA	1-800-352-8763
International Chimney Corp.	Huntsville AL	1-800-828-1446
International Technology Corp.	Torrance CA	1-800-421-5574
JACA Corp.	Fort Washington PA	1-800-292-2510
Jaeger Products Inc.	Houston TX	1-800-678-0345
Kimble/Kontes	Vineland NJ	1-800-223-7150
Koger/Air Corp.	Martinsville VA	1-800-368-2096
Lab Safety Supply Co.	Janesaville WI	1-800-356-0783
Laser Science Inc.	Newton MA	1-800-447-1020
Lear Siegler Measurement Controls Corp.	Englewood CO	1-800-422-1499
Longyear US Products Group	Stone Mountain GA	1-800-241-9468
MAC Environmental	Sabetha KS	1-800-223-2191
MAC Equipment Inc.	Kansas City MO	1-800-821-2476
MDA Scientific Inc.	Lincolnshire Il	1-800-344-4632
Metal Reaction Inc.	Hialeah FL	1-800-826-8856
Metco Environmental	Addison TX	1-800-394-1194
Mid-Atlantic Environmental Inc.	Broomall PA	1-800-742-7687
Mojave Desert Air Quality Management District-Complaint Line		1-800-635-4617
Monroe Environmental Corp.	Monroe MI	1-800-992-7707
MSA	Pittsburgh PA	1-800-MSA2222
Munters Corp.	Ft Myers FL	1-800-446-6868
Nao Inc.	Philadelphia PA	1-800-523-3495
National Draeger Inc.	Pittsburgh PA	1-800-922-5518
North American Weather Consultants	Salt Lake City UT	1-800-658-8493
Omega Engineering Inc.	Stamford CT	1-800-826-6342

ORS Environmental Equipment	Greenville NH	1-800-228-2310
Osmonics	Minnetonka MN	1-800-848-1750
Perkin Elmer	Norwalk CT	1-800-762-4000
PrecipTech/BHA Group	Kansas City MO	1-800-336-2585
Protrans International Inc.	Randolph MA	1-800-826-0807
Purified Microenvironments Inc.	Miami FL	1-800-888-5357
QED Environmental Systems Inc.	Ann Arbor MI	1-800-624-2026
Reach Associates	S Orange NJ	1-800-246-9628
Rosemount Analytical Inc.	Santa Clara CA	1-800-628-1200
Ross Incineration Services Inc.	Grafton OH	1-800-878-7677
Rusmar Inc.	West Chester PA	1-800-733-3626
Rust Environmental & Infrastructure	Greensville SC	1-800-868-0373
Sensidyne Inc.	Clearwater FL	1-800-451-9444
Servomex Co.	Norwood MA	1-800-862-0200
SHENA Inc.	Norcross GA	1-800-25SHENA
Siemens Industrial Automation Inc.	Alpharaetta GA	1-800-964-4114
SKC Inc.	Eighty-Four PA	1-800-752-8472
Sly Inc./W.W. Sly Manufacturing Co.	Strongsville OH	1-800-334-2957
Solvay Interox	Houston TX	1-800-INTEROX
Special Resource Management	Butte MT	1-800-735-8964
Spectronics	Westbury NY	1-800-274-8888
Standard Filter Corp.	Carlsbad CA	1-800-634-5837
Statgraphics/Manugistics	Rockville MD	1-800-592-0050
Summit Filter Corp.	Union NJ	1-800-321-4850
Sunshine Instruments	Philadelphia PA	1-800-343-1199
Tectronic Products	E Syracuse NY	1-800-227-1375
Tepco Air Pollution Control Services	Beverly MA	1-800-675-2927
Torit & Day, Donaldson Co. Inc.	Minneapolis MN	1-800-365-1331
Ultra Industries Inc.	Bellwood IL	1-800-358-5872
United Air Specialists Inc.	Cincinnati OH	1-800-922-4422
Universal Air Precipitator Corp.	Monroeville PA	1-800-326-8406
VIC Environmental Systems Inc.	Minneapolis MN	1-800-669-8777
Wahlco	Thornton IL	1-800-882-8100
Wedding & Assoc. Inc.	Ft Collins CO	1-800-367-7610
Westport Environmental Systems	Westport MA	1-800-343-9411
Wheelabrator Environmental Systems Inc.	Hampton NH	1-800-682-0026

ASBESTOS

A A C M Co	Louisville KY	1-800-228-2502
AAAA by Phillips	Bristol PA	1-800-445-4166
AAR Contractor Inc	Albany NY	1-800-242-0753
Aaron Environmental Specialists	Waterbury CT	1-800-248-9858
AB Control	Raleigh NC	1-800-332-0835
Abatement Services	Shreveport LA	1-800-544-3543
ACM Management Inc.	Del City OK	1-800-262-1461
Aerospace America	Bay City MI	1-800-843-6168
Affordable Abatement	Woodinville WA	1-800-545-9411
All South Environmental Services	Kingsport TN	1-800-533-2989
All South Environmental Services	Kingsport TN	1-800-533-2989
All Temp Inc	Kansas City MO	1-800-874-0028
Allwash Of Syracuse	Syracuse NY	1-800-633-9274
Allwash of Syracuse	Syracuse NY	1-800-633-9274
Allwaste Asbestos Abatement	Casselberry FL	1-800-535-6878
Allwaste Asbestos Abatement	Houston TX	1-800-535-6878
American Asbestos Removal	Detroit MI	1-800-962-7813
American Coatings Corp	Pompano Beach FL	1-800-533-0151

American Coatings Corp	Pompano Beach FL	1-800-323-7580
American Technologies	Orange CA	1-800-400-9353
Amerisafe	Melrose Park IL	1-800-433-2297
Applied Environmental Science Inc.	Minneapolis MN	1-800-626-8089
ARC Inc.	Tuscaloosa AL	1-800-451-0047
Arpro	Temple PA	1-800-992-7776
Asbat Services	Columbus OH	1-800-852-9521
Asbestos Analytics	Moore OK	1-800-343-0517
Asbestos Analytics	Corona CA	1-800-832-1995
Asbestos Control Inc.		1-800-322-1089
Asbestos Environmental Control Service	Cockeysville MD	1-800-423-0149
Asbestos Removal	Manasquan NJ	1-800-331-2547
Asbestos Removal	Madison WI	1-800-835-6047
Asbestos Removal Technology	Vancouver WA	1-800-321-4121
Asbestos Safety Technologies of Califor	Santee CA	1-800-742-5365
Asbestos Technical Information Service	Research Triangle Park NC	1-800-334-8571
Asbestos Technology & Consulting	Colorado Springs CO	1-800-732-7670
Atlantic Abatement	Providence RI	1-800-962-2283
B & R Insulation	Lenexa KS	1-800-462-9462
B D C Services Inc	Sacramento CA	1-800-228-4232
Balsam Environmental Consultants Inc.	Salem NH	1-800-933-9322
Baratta & Associates	Vinton VA	1-800-543-7807
Barsottis Inc.	Norfolk VA	1-800-252-5744
BDC Services	Azusa CA	1-800-642-4232
Best Construction Temporaries & Trainin	Greensboro NC	1-800-325-2378
Brand Northwest Inc.	Kent WA	1-800-562-1333
Brunswick Construction	Waterford NY	1-800-562-6433
California Asbestos Monafilla Inc	Copperopolis CA	1-800-852-4031
Casa Construction Ltd.	Sierra Vista AZ	1-800-344-0001
Chempower Inc.	N Canton OH	1-800-422-4299
Clark Tech Environmental Systems	Houston TX	1-800-245-0509
Colorado Environmental Control Division	Commerce City CO	1-800-732-3060
Control Resource System Inc.	Mars PA	1-800-445-2774
Control Resource System Inc.	Kent WA	1-800-845-4547
Control Resource Systems Inc.	Lakeland FL	1-800-662-1385
Control Resources Systems Inc.	Chandler AZ	1-800-548-0133
Critical Services	Kent WA	1-800-624-7030
Cyclone Steeple Jacks	Nevada IA	1-800-282-3123
Daley Insulation Co. Inc.	Winsted CT	1-800-826-7029
Dawson Associates	Lawrenceville GA	1-800-282-4782
Dawson Associates	Lawrenceville GA	1-800-282-4782
Dec-Tam	Andover MA	1-800-332-8261
Del-Tech Environmental	Citrus Heights CA	1-800-547-4921
Delta O Services	Austin TX	1-800-634-9362
Dore & Associates Contracting	Bay City MI	1-800-344-7876
Dore & Associates Contracting	Bay City MI	1-800-344-7876
DSR Asbestos Removal Co. Inc.	Staten Island NY	1-800-424-2377
E & W Construction and Asbestos Abatement		1-800-572-1123
Eneco Tech Inc.	Denver CO	1-800-659-TECH
Enviromental Action	Depew OK	1-800-643-5724
Environet of Wisconsin	Winneconne WI	1-800-542-8940
Environment Technology Of Fort Wayne	Ft Wayne IN	1-800-448-1778
Environmental Control Division	Commerce City CO	1-800-732-3060
Environmental Control Division	Dallas TX	1-800-368-8371
Environmental Hazard Control	Wauconda IL	1-800-543-1499
Environmental Inovations	Oakland CA	1-800-554-0037
Environmental Resource Associates Of Florida Inc		1-800-221-3486
Enviropore Inc	Bala-Cynwyd PA	1-800-874-6270

Enviropore Inc.	Lumberton NJ	1-800-874-6270
EWT Contracting	New York NY	1-800-242-4598
Fiberlock Technology	Cambridge MA	1-800-342-3755
Flexi-Wall	Liberty SC	1-800-843-9318
Fortress Industries	De Witt MI	1-800-526-2569
Gem Environmental	Williamstown MA	1-800-992-3533
Gerry Hobson General Contracting Inc.	Brush Prairie WA	1-800-848-4160
Global Consumer Services	Los Angeles CA	1-800-233-6228
Global Consumer Services	Los Angeles CA	1-800-233-6226
Glynagin G P Enterprises	Covington LA	1-800-256-4457
Great Plain Asbsetors Control	Kearney NE	1-800-445-0067
Green Mountain Environmental Services	Essex VT	1-800-437-4136
High Valley Environmental	Denver CO	1-800-524-5805
High Valley Environmental	Denver CO	1-800-524-5805
His 3 Abatement	York PA	1-800-758-4733
Holian Asbestos Removal	Spring Grove IL	1-800-626-7566
Hoosier Abatement	Indianapolis IN	1-800-282-5332
HzW Environmental Consultants	Mentor OH	1-800CONSULT
I & F	Cincinnati OH	1-800-922-2569
Ideal and Assoc Environmental Engineeri	Bloomington IL	1-800-535-0964
Inflatable Abatement Systems	E Millinocket ME	1-800-962-2565
Infosafe Inc Sales	Broomfield CO	1-800-344-8839
Infosafe Inc.	Westminster CO	1-800-344-8839
Insul-Coat	Orlando FL	1-800-437-5768
Insulated Systems	Erie PA	1-800-851-2391
Insulated Systems	Erie PA	1-800-851-2391
Insulation Removal Corp	Boring OR	1-800-548-6606
Insulation Specialists	Hillsboro OR	1-800-234-6105
Insulation Technologies Inc	Harvey LA	1-800-762-1550
Insulation Technologies Inc.	Harvey LA	1-800-762-1550
Interstate Industrial Mechanical Inc.	Stevenson WA	1-800-637-4236
Iowa Illinois Thermal Insulation	Davenport IA	1-800-292-1280
IT Corp.	Wilmington CA	1-800-421-5574
J & D Enterprises of Duluth	Duluth MN	1-800-541-7511
Jim Macklin	Riverside CA	1-800-845-7518
Keeson Inc.	Phoenix AZ	1-800-543-2522
Laidlaw Environmental Services Inc.	Columbia SC	1-800-356-8570
Laughlin Development	Houston TX	1-800-344-3141
Leader Industries	Portage IN	1-800-437-6122
Lepi Enterprises	Zanesville OH	1-800-327-5374
LVI Environmental Services Inc.	San Leandro CA	1-800-343-2372
Makson Corp	Patterson NJ	1-800-831-4335
Mello Mfg	Richmond CA	1-800-342-7776
Milwaukee Asbestos Information Center	Milwaukee WI	1-800-848-3298
Milwaukee Asbestos Information Center	Milwaukee WI	1-800-848-3298
Miram Construction	W Paterson NJ	1-800-235-6472
MJM Asbestos Abatement	Melrose MA	1-800-339-3447
MK Moore and Sons	Dayton OH	1-800-542-6598
Mystic Air Quality Consultants	Groton CT	1-800-247-7746
National Abatement Services Inc.	Hingham MA	1-800-622-6060
National Environmental	Oakland CA	1-800-922-9984
National Surface Cleaning Inc	Methuen MA	1-800-962-5110
National Surface Cleaning Inc.	Elmwood Park NJ	1-800-962-5110
NEO	Hazelwood NC	1-800-822-1247
New England Abatement Resources Inc.	Canton MA	1-800-421-6327
OHM	Clarence Center NY	1-800-457-4412
OHM Corp.	Findlay OH	1-800-537-9540
OHM Environmental	Lanham MD	1-800-662-7618

Oilind Safety	Denver CO	1-800-843-4855
P N S Jewelry	Los Angeles CA	1-800-647-0675
Pacific Environmental Control	San Ramon CA	1-800-822-8489
Phoenix Disposal	Garden City NY	1-800-635-0284
Potomac Abatement	Baltimore MD	1-800-328-8860
Precision Environmental Co	Independence OH	1-800-553-5089
Precision Environmental Co	Independence OH	1-800-553-5089
Princeton Testing Laboratory	Princeton NJ	1-800-548-TEST
Professional Services	Lawrence KS	1-800-445-0682
PW Stephens	Reno NV	1-800-551-8387
PW Stephens Contractors	Reno NV	1-800-551-8228
Reed & Greenwood Insulation Inc.	Farmington CT	1-800-842-9056
Reliance Environmental Management	Ft Walton Beach FL	1-800-822-7525
REM Con Inc.	Coon Rapids MN	1-800-343-6182
Removal and Abatement Tech	Augusta GA	1-800-235-4108
Risk Removal	Ft Collins CO	1-800-322-9121
RL McClean Asbestos Abatement	Wallingford CT	1-800-228-5789
RNK Safety Supplies	Red Oak TX	1-800-628-8036
RNK Safety Supplies	Red Oak TX	1-800-628-8036
Rust Environmental & Infrastructure	Concord NH	1-800-868-0373
Safeair Systems	Minneapolis MN	1-800-223-1194
Safety Environmental Control Inc	E Swanzey NH	1-800-543-4592
Scott Allard & Bohannon	Phoenix AZ	1-800-253-0044
Seneca Asbestos Removal & Control	Tiffin OH	1-800-752-1557
Seneca Asbestos Removal & Control	Middletown OH	1-800-334-9622
Sierra Environmental Group Inc.	Blacklick OH	1-800-892-5988
Simpson and Associates Inc.	Trilby FL	1-800-222-9898
Smith Supply	Temple TX	1-800-242-8975
Southern Concepts	Hyattsville MD	1-800-683-3050
Specialty Vacuum Inc	Valley Park MO	1-800-448-2801
Spirco Environmental	St Louis MO	1-800-462-7743
Stephens P W Contractors	Roseville CA	1-800-551-8228
Sverdup Civil Inc.	Maryland Heights MO	1-800-325-7910
TBN Assoc	Beltsville MD	1-800-876-9745
The Brand Companies	Westchester IL	1-800-468-1480
The ERM Group	Exton PA	1-800-544-3117
The Group Solution Inc.	Salem NH	1-800-781-0051
Tom House Insulation & Asbestos Management Systems		1-800-722-3078
TRC Environmental Consultants	E Hartford CT	1-800-367-1044
TRC Environmental Corp.	Windsdor CT	1-800-TRC-5601
U S A Asbestos	Clifton NJ	1-800-523-3446
USA Remediation Services Inc	Gainesville VA	1-800-531-7392
Webster Environmental	Ruffin NC	1-800-526-1041
Wyoming S & P	Wilkes-Barre PA	1-800-232-6033

ASSOCIATIONS

ABC Girl Scouts	Richmond VA	1-800-447-5728
African Wildlife Foundation	Washington DC	1-800-344-8875
Alliance For Community Trees	Sacramento CA	1-800-228-8886
Alliance For Quality Education	Dallas TX	1-800-282-8080
American Assoc Of University Professors	Washington DC	1-800-424-2973
American Birding Assoc	Colorado Springs CO	1-800-835-2473
American Cancer Society		1-800-227-2345
American Chemical Society	Washington DC	1-800-227-5558
American College Of Allergy & Immunolog	Palatine IL	1-800-842-7777
American Discovery Trail	Virginia Beach VA	1-800-851-3442

American Forage & Grassland	Georgetown TX	1-800-944-2342
American Forestry Association	Washington DC	1-800-368-5748
American Horticultural Society	Alexandria VA	1-800-777-7931
American Horticultural Therapy Association		1-800-634-1603
American Hospital Association		1-800-242-2626
American Institute of Architects	Denver CO	1-800-628-5598
American Institute of Architects	Phoenix AZ	1-800-367-2781
American Institute of Biological Scienc	Washington DC	1-800-992-2427
American Institute of Chemical Engineer	New York NY	1-800-242-4363
American Nuclear Society	La Grange IL	1-800-682-6397
American Plastics Council Hotline		1-800-2-HELP90
American Roentgen Ray Society Of Reston	Reston VA	1-800-438-2777
American Small Businesses Assoc	Arlington VA	1-800-235-3298
American Society For Quality Control	Milwaukee WI	1-800-248-1946
American Society Of Mechanical Engineer	Mt Prospect IL	1-800-221-5536
American Society Of Mechanical Engineer	Mt Prospect IL	1-800-445-2388
American Society Of Mechanical Engineer	Mt Prospect IL	1-800-628-5981
American Society Of Mechanical Engineer	Mt Prospect IL	1-800-628-6437
American Society of Mechanical Engineer	Eastern	1-800-221-5536
American Society of Mechanical Engineer	Midwest	1-800-628-6437
American Society of Mechanical Engineer	Northeast	1-800-628-5981
American Society of Mechanical Engineer	Southern	1-800-445-2388
American Society of Mechanical Engineer	Western	1-800-624-9002
American Society of Mechanical Engineer	Denver CO	1-800-926-7337
American Society Of Mechanical Engineer	San Rafael CA	1-800-624-9002
American Teachers Assoc	Austin TX	1-800-832-2488
American Therapeutic Recreation Assoc	Hattiesburg MS	1-800-553-0304
American Water Works Association	Miami FL	1-800-443-9353
American Welding Society	McLean VA	1-800-524-0475
AMT The Assoc For Mfg Technology	McLean VA	1-800-524-0475
Appalachian Girl Scout Council	Hamilton MT	1-800-775-3244
ARC-National Employment & Training Prog	Winchester MA	1-800-833-2627
Arctic Grayling Recovery Program	Las Vegas NV	1-800-422-1771
Association for Manufacturing Technolog	Johnson City TN	1-800-428-3991
Association Management Consulting Resou	Brigham UT	1-800-445-4227
Association Of Contingency Planners	Ogden UT	1-800-445-4227
Association of Contingency Planners		1-800-438-8247
Boy Scouts Of America	Columbus OH	1-800-433-4051
Boy Scouts Of America Arkansas	Little Rock AR	1-800-545-7268
Boy Scouts of America Central Ohio Coun	Dayton OH	1-800-233-4845
Buckeye Trails Girl Scout Council	St. Petersburg FL	1-800-262-3567
Center for Marine Conservation	Richmond VA	1-800-352-4390
Central Virginia Safety Council	Richmond VA	1-800-262-7905
Central Virginia Safety Council	Bloomington MN	1-800-445-1525
Citizen Band Potawatomi Tribe	Seattle WA	1-800-722-8055
Communicating for Agriculture	Bloomington MN	1-800-523-5257
Communicating for Health Consumers	San Francisco CA	1-800-652-1080
Dinosaur Society The	E Islip NY	1-800-346-6366
Earth Day	New York NY	1-800-332-7843
Electric and Gas Industries Association	Northfield IL	1-800-732-8878
Endangered Species Federation	Marlboro NJ	1-800-327-3634
Environmental Chemical Association	Washington DC	1-800-225-5333
Environmental Defense Fund Hotline	Seattle WA	1-800-654-0274
Environmental Industry Association	Washington DC	1-800-424-2869
Flint River Girl Scout Council	Albany GA	1-800-448-4762
Flordia Society of Professional Land Su	Tallahassee FL	1-800-237-4384
Food for Thought	Brooklyn NY	1-800-647-7780
Foundation For Safe Boating Marine Info	Brooklyn NY	1-800-647-7780

Foundation for Safe Boating Marine Info	San Francisco CA	1-800-421-9283
Girl Scouts Joshua Tree Council	Bakersfield CA	1-800-225-4475
Girl Scouts Pacific Peak Council	Tumwater WA	1-800-541-9852
Girl Scouts Pacific Peak Council	Tacoma WA	1-800-874-5260
Girl Scouts Pisgah Council	Asheville NX	1-800-522-6280
Girl Scouts Plymouth Bay Council	Taunton MA	1-800-242-0925
Great Whales Foundation	Washington DC	1-800-634-1598
Hazardous Materials Advisory Council	Washington DC	1-800-634-1598
Hazardous Materials Advisory Council	Mount Prospect IL	1-800-634-0911
Illinois Fire Safety Alliance	Springfield IL	1-800-321-5422
Illinois Land Improvement	Odessa TX	1-800-592-1433
Industrial Foundation of America	Nashville TN	1-800-952-3223
Kids for a Clean Environment	Los Angeles CA	1-800-527-7013
Los Angeles Radiological Society	Ambler PA	1-800-851-1145
Metropolitan Philadlphia Safety & Healt	Michigan Center MI	1-800-242-4638
Michigan Fire Safety Foundation	Lakewood CO	1-800-526-7943
Mountain States Association	Reno NV	1-800-344BUCK
National Association of Environmental P	Morganville NJ	1-800-707-1578
National Association of Socially Respon	Columbus OH	1-800-551-7379
National Ground Water Association	Kennesaw GA	1-800228NOOW
National Organization of Outdoor Women	Washington DC	1-800-448-6722
National Parks & Conservation	Duluth MN	1-800-562-0004
National Resource Center	Englewood CO	1-800-227-5242
National Rifle Association	Fairfax VA	1-800-672-3888
National Safety Council	Redwood City CA	1-800-441-5103
National Safety Council	Syracuse NY	1-800-621-7615
National Safety Council	Redwood City CA	1-800-822-7009
National Safety Council	Redwood City CA	1-800-962-3434
National Safety Council	Weaverville NC	1-800-321-2910
National Safety Management Society	Weaverville NC	1-800-321-2910
National Safety Management Society	Washington DC	1-800-535-7148
National Society of Consulting Soil Sci	Washington DC	1-800-672-4183
National Solid Waste Management Assoc	Naperville IL	1-800-679-6269
National Trails Day	Vienna VA	1-800-972-8608
National Transportation Assoc	Rowlett TX	1-800-595-1995
National Trust for Historic Preservatio	Wilmington NC	1-800-457-9700
National Vocational Agriculture Teacher	Alexandria VA	1-800-772-0939
National Wild Turkey Association	Edgefield SC	1-800-843-6983
National Wild Turkey Federation	Grand Island NE	1-800-248-7328
National Wildlife Fed	Vienna VA	1-800-432-6564
National Wildlife Federation	Vienna VA	1-800-822-9919
Natural Resources Research Institute	Itasca IL	1-800-621-7619
Nebraska State Recycling Association	Minnetonka MN	1-800-843-6232
NJ Environmental Protection Department	Trenton NJ	1-800-648-7263
North American Fishing Club	Minnetonka MN	1-800-843-6232
North American Fishing Club	Minnetonka MN	1-800-922-4868
North American Hunting Club	Minnetonka MN	1-800-922-4868
North American Hunting Club	Meredith NH	1-800-462-5666
North American Loon Fund	Washington DC	1-800-456-6323
Ocean Club	San Antonio TX	1-800-262-2282
Oregon Assoc Of Nurserymen	Milwaukie OR	1-800-342-6401
Oregon Association of Nurseryman	Kihei HI	1-800-942-5311
Organization for Economic Cooperation	Milwaukie OR	1-800-342-6401
Pacific Whale Foundation	University Park PA	1-800-235-9473
Pennsylvania Forest Stewardship Program	University Park PA	1-800-235-9473
Pennsylvania Forest Stewardship Program	New York NY	1-800-822-3020
Portable Sanitation Assoc	Bloomington MN	1-800-822-3020
Portable Sanitation Association		1-800-788-7464

1-800-NAT PARK

Potawatomi Indian Tribe	Irving TX	1-800-742-3075
Professional Environmental Trainers Ass	Old Hickory TN	1-800-232-4537
Professional Plant Growers Assoc	Mason MI	1-800-647-7742
Protect American Eagles	Federalsburg MD	1-800-888-7747
Rails to Trails Conservancy		1-800-460-RAIN
Rainforest Preservation Foundation	Miami FL	1-800-437-3532
Releaseadermy	Missoula MT	1-800-843-7633
Rocky Mountain Elk Foundation	Missoula MT	1-800-843-7633
Rocky Mountain Elk Foundation		1-800-432-JOIN
Sault Ste Marie Tribe Of Chippewa India	Sault Ste Marie MI	1-800-251-6597
Save the Florida Panther Inc	St Petersburg FL	1-800-535-2228
Save The Manatee		1-800-859-SAVE
Save the Manatee Club	Maitland FL	1-800-432-5646
Sea Turtle Restoration Project		1-800-678-7853
Sea Turtle Survival League	Washington DC	1-800-457-4474
Solid Waste Composting Council	Tallahassee FL	1-800-441-7949
Southern Waste Information Exchange		1-800-ASK-FISH
Sportfishing Promotion Council	Austin TX	1-800-322-8492
Texas Game Warden Assoc	Austin TX	1-800-322-8492
Texas Game Warden Association	Washington DC	1-800-243-5790
The Billfish Foundation	Columbus OH	1-800-433-4051
The Council for Solid Waste Solutions	Washington DC	1-800-727-9618
The Evergreen State Society	Lakeland FL	1-800-433-3971
The Mule Deer Foundation	Washington DC	1-800-526-6237
Times Mirror Magazines Conservation Cou	New York NY	1-800-533-2784
Trees For The Future	Lacey WA	1-800-533-0852
Trout Unlimited	Richmond VA	1-800-633-2208
United Agribusiness League	Irvine CA	1-800-223-4590
United Sportsmens Assoc	Colorado Springs CO	1-800-872-5207
Urban Land Institute	Arnold MD	1-800-321-5011
US Kickapoo Tribe Social Services	Horton KS	1-800-732-9414
US Naval Institute	Annapolis MD	1-800-233-8764
Virginia Safety Assoc	Richmond VA	1-800-633-2208
Virginia Safety Association	Boston MA	1-800-554-3569
Walden Walk A Thon	Boston MA	1-800-543-9911
Walden Woods Project	Alexandria VA	1-800-556-3700
Washington State Pest Control Assoc	Belfair WA	1-800-253-3836
Water Pollution Control Federation	Cherry Hill NJ	1-800-245-9772
Western Pennsylvania Safety Council	Madison WI	1-800-344-0421
Wilderness Heritage Society	Austin TX	1-800-588-5559
Wisconsin Land Information Assoc	Wildrose WI	1-800-344-0421
Wisconsin Waste Carrier Association	Washington DC	1-800-634-4444
World Wildlife Fund		1-800-4-YIPPEE

BINOCULARS, RANGEFINDERS & NIGHTVISION

AGS Company	San Francisco CA	1-800-540-1000
Bausch & Lomb Sports Optics	Overland Park KS	1-800-423-3537
Binocular City		1-800-473-1621
Carl Zeiss Optical	Petersburg VA	1-800-338-2984
Celestron International	Torrance CA	1-800-421-1526
Compass Industries Inc.	New York NY	1-800-221-9904
Europtik Ltd.	Dunmore PA	1-800-873-5362
Fujinon Optical	Wayne NJ	1-800-872-0196
HGS Outfitters		1-800-545-0152
HiTek	Redwood City CA	1-800-54-NIGHT
Innovative Weaponry Inc.	Albuquerque NM	1-800-334-3573
Joy Enterprises	Ridgefield NJ	1-800-526-0486

Leica Camera Inc.	Northvale NJ	1-800-222-0118
Monument Camera & Video	Boise ID	1-800-777-8550
National Camera Exchange		1-800-624-8107
Olympic Optical Co.	Memphis TN	1-800-238-7120
Optolyth - USA	Hillsboro OR	1-800-447-6881
Optronics	Ft Gibson OK	1-800-364-LITE
Ranging	E Bloomfield NY	1-800-724-7486
Rosenberg Jack & Sons	Dallas TX	1-800-527-7537
Sagittarius	San Diego CA	1-800-999-2114
SCM Corp		1-800-225-9407
Steiner/Pioneer Research Inc.	Westmont NJ	1-800-257-7742
Swarovski Optik	Cranston RI	1-800-426-3089
Swift Instrument Inc.	Boston MA	1-800-446-1116

CHEMICAL CLEANING - INDUSTRIAL

Ace Soap Factory	San Antonio TX	1-800-373-3970
Autosport Inc	Bound Brook NJ	1-800-355-3505
Chemaster	Midville GA	1-800-643-8455
Chemtronics Inc.	Kennesaw GA	1-800-645-5244
Copens Powerline Industrial Center	Pompano Beach FL	1-800-545-2306
D S Chemical Specialties	Portland OR	1-800-769-2436
Detergent Mktg Systems	Rancho Cucamonga CA	1-800-924-2842
Environmental Quik Pak	Lynbrook NY	1-800-284-5584
Griffin Tommy	Houston TX	1-800-362-2902
Hocking Intl	National City CA	1-800-533-8151
Mandate Inc	Atlanta GA	1-800-626-3283
Saroha Inc	Baytown TX	1-800-717-2002
Tech Spray Inc	Amarillo TX	1-800-858-4043
Texas National Chemical Company Inc.	Dallas TX	1-800-443-2462
Unisource Industries	Chicago IL	1-800-822-7766
United Lab Inc	Bellingham WA	1-800-858-3558

CHEMICALS

5-C Chemical	Lufkin TX	1-800-752-2436
A & V Inc	Sussex WI	1-800-833-2334
A Global Chemical	Arlington TX	1-800-433-5183
A&L Labs	Minneapolis MN	1-800-225-3832
ABC Chemical Corp	Detroit MI	1-800-448-2089
AC West Virginia Polyols	Birmingham MI	1-800-233-6461
ACSI	Milpitas CA	1-800-831-2274
Aakash Chemicals & Dyestuffs	Addison IL	1-800-255-4855
Abbott Lab Agricultural Chemical Divisi	Memphis TN	1-800-535-6545
Accurate Chemical & Scientific	San Diego CA	1-800-255-9378
Acme Hardesty	Jenkintown PA	1-800-223-7054
Acton Technologies Inc	Pittston PA	1-800-654-0612
Adams County Co-Op	Hoagland IN	1-800-292-7270
Adapco	Sanford FL	1-800-367-0659
Aegis Floor Systems Southeast	Sand Springs OK	1-800-752-1425
Aetna Chemical	Elmwood Park NJ	1-800-345-2518
Air Products And Chemicals	Granite City IL	1-800-851-8021
Airco Industrial Gases	Middlesex NJ	1-800-356-0257
Akzo Chemical Division	Chicago IL	1-800-227-7070
Akzo Chemical Division	Chicago IL	1-800-828-7929
Akzo Chemical Inc	Chicago IL	1-800-257-8292
Akzo Chemical Inc	Covina CA	1-800-426-2506
Akzo Coatings & Resin Operation	Louisville KY	1-800-292-2308

Akzo Dreeland	Denver CO	1-800-662-8170
Akzo Nobel	Dobbs Ferry NY	1-800-666-1200
Al Lifson & Associates	Forked River NJ	1-800-624-8088
Alcan Chemicals	Beachwood OH	1-800-321-3864
Alco Chemical Corp	Chattanooga TN	1-800-251-1080
Aldrich Chemical Technical Services	Milwaukee WI	1-800-558-2436
Aldrich Chemical	Milwaukee WI	1-800-227-4563
Aldrich Chemical	Milwaukee WI	1-800-231-8327
Aldrich Chemical	Milwaukee WI	1-800-255-3756
Aldrich Chemical	Milwaukee WI	1-800-336-9719
Aldrich Chemical	Milwaukee WI	1-800-558-9160
Aldrich Chemical	Milwaukee WI	1-800-771-6737
Alexander Chemical	Lisle IL	1-800-445-9458
Alexander Chemical	Laporte IN	1-800-468-5144
All Kem Co	Live Oak FL	1-800-814-6447
All Parts Distribution	Indianapolis IN	1-800-428-4731
All Star Chemicals	San Diego CA	1-800-362-5031
Allied Kelite Division Witco Chemical	Los Angeles CA	1-800-223-4520
Allied Kelite	Los Angeles CA	1-800-243-9927
Amano Intl Enzyme	Troy VA	1-800-446-7652
Amcor	Medina OH	1-800-227-3574
Amerchol	Edison NJ	1-800-367-3534
Ameribrom West	Camarillo CA	1-800-845-6661
American Building Restoration Inc	Franklin WI	1-800-346-7532
American Chemical Diversified	Pelham NY	1-800-782-2804
American Chemical Inc	St Louis MO	1-800-633-0247
American Chrome & Chemical	Corpus Christi TX	1-800-242-3075
American Chrome & Chemical	Corpus Christi TX	1-800-531-3188
American Cyanamid	S Charleston WV	1-800-826-2544
American Gilsonite	Salt Lake City UT	1-800-457-6483
American Niagara	Atlanta GA	1-800-241-7708
American Radio Labeled Chemicals	St Louis MO	1-800-331-6661
American Technologies, Inc.	Novi MI	1-800-338-3743
Amoco Chemicals	Chicago IL	1-800-621-8888
Amrep	Marietta GA	1-800-221-6772
Anchor Americhem	Cumming GA	1-800-243-8010
Ankem Chemicals	Dallas TX	1-800-527-7615
Apex Engineering Products	Plainfield IL	1-800-451-6291
Apollo Industries	Smyrna GA	1-800-533-3548
Aqua Process Inc	Houston TX	1-800-323-1435
Aqualon Technial Mktg	Wilmington DE	1-800-345-0447
Aqualon	Huntington Beach CA	1-800-272-0262
Archem	Portsmouth OH	1-800-635-1125
Arco Chemical	Hinsdale IL	1-800-321-7000
Arco Chemical	Hinsdale IL	1-800-354-1550
Argent Chemical Laboratory	Redmond WA	1-800-426-6258
Argus Division Of Witco Corp	Charlotte NC	1-800-325-2481
Aristech Chemical Corp	Jacksonville AR	1-800-643-5592
Aristech Chemical	Medley FL	1-800-432-8971
Arizona Chemical Co Panama City Fl	Panama City FL	1-800-526-5294
Armand Products	Princeton NJ	1-800-522-0540
Armand Products	Princeton NJ	1-800-526-3563
Arol Chemical Products	Newark NJ	1-800-783-2765
Arrochem	Mt Holly NC	1-800-438-5883
Arrow Chemical Corp	Lynbrook NY	1-800-282-7769
Ashland Chemical Division Ashland Oil	Doraville GA	1-800-282-8910
Ashland Chemical IC & S	Tampa FL	1-800-237-6283
Ashland ICCS Division	Aston PA	1-800-523-1738

Associated Chemists	Milwaukie OR	1-800-554-4666
Associated Products	Glenshaw PA	1-800-243-5689
Astro Chemicals	Springfield MA	1-800-959-7240
Astro Chemical	Springfield MA	1-800-223-0776
Astro Industries	Morganton NC	1-800-872-7876
Atlantic Metals	Philadelphia PA	1-800-352-1201
Atlas Chemical Inc	Portland OR	1-800-272-8527
Atochem North America Inc-Decco	McAllen TX	1-800-553-6756
Avitrol Corp	Tulsa OK	1-800-633-5069
Ayers Intl Corp	Irvington NY	1-800-852-8396
Azok Group The	Lorain OH	1-800-730-3904
B & A Chemical	Gardner KS	1-800-235-9914
BASF Corp Chemicals Division	Wyandotte MI	1-800-521-9100
BASF Corp-Pigments	Holland MI	1-800-433-4731
BASF Corp/ C & I Division	Miami Township OH	1-800-543-7168
BASF Corp	Houston TX	1-800-227-3839
BD Chemicals Specialty Chemicals	Salinas CA	1-800-662-0378
BP Chemicals Americas Inc	Hackettstown NJ	1-800-272-4367
BTL Specialty Resins Corp	Blue Island IL	1-800-451-5707
Baerlocher USA	Dover OH	1-800-342-6158
Baker Performance Chemicals	Houston TX	1-800-231-3606
Baker Performance Chemical	Laurel MS	1-800-521-0941
Baker Performance-Aquaness	Bakersfield CA	1-800-847-5793
Baker Pipeline Products	Houston TX	1-800-225-3793
Baron Lab	Dallas TX	1-800-552-6442
Becker Underwood	Ames IA	1-800-232-5907
Benchmark	Wyandotte MI	1-800-521-9107
Bengal Chemical	Baton Rouge LA	1-800-367-0394
Berkshire Color & Chemical Corp	E Longmeadow MA	1-800-336-8558
Biopharm Inc	Hatfield AR	1-800-443-8465
Biosynth Intl	Skokie IL	1-800-270-2436
Biozyme Lab Intl Ltd	San Diego CA	1-800-423-8199
Blue Grass Chemical Specialties	New Albany IN	1-800-638-7197
Boehringer Mannheim Biochemicals	Indianapolis IN	1-800-428-5433
Boliden Intertrade Inc	Atlanta GA	1-800-241-1912
Borchers Supply	Hawarden IA	1-800-522-4731
Borden & Remington	Fall River MA	1-800-543-5393
Boysenblue Intl	Lafayette LA	1-800-624-2146
Brach Chemical	Miami FL	1-800-889-9980
Break Thru	Hayward CA	1-800-323-2215
Brin-Mont Chemicals	Greensboro NC	1-800-662-1261
Buckman Laboratories	Memphis TN	1-800-282-5626
Buffalo Color Corp	Parsippany NJ	1-800-631-0171
Burdick & Jackson	Muskegon MI	1-800-368-0050
C H Patrick & Co Inc	Greenville SC	1-800-247-2586
C R C Industries Inc	Warminster PA	1-800-272-4560
C R C Industries Inc	Warminster PA	1-800-556-5074
C V D Systems & Services Inc	Quakertown PA	1-800-544-6498
CC Oil & Chemical	Cleveland OH	1-800-321-4202
CWC Industries	Cleveland OH	1-800-292-5755
Cal Chlor Inc	Opelousas LA	1-800-245-6743
Calcium Chloride Sales Inc	Grove City PA	1-800-228-3879
Calgene Chemicals Inc	Skokie IL	1-800-432-7187
Calgon Vestal Credit	St Louis MO	1-800-558-0098
Calgon Vestal Lab	St Louis MO	1-800-325-8005
Calgon Vestal Phone Mail	St Louis MO	1-800-523-1735
Calgon Vestal-Fax Line	St Louis MO	1-800-543-2680
Callahan Chemical	Palmyra NJ	1-800-257-7967

Cambridge Isotope Labs	Woburn MA	1-800-322-1174
Canyon Industries	Casa Grande AZ	1-800-255-3423
Cape Industries DMT PTA Products	Wilmington NC	1-800-441-2584
Cape Industries	Wilmington NC	1-800-648-2717
Capehorn Methanol USA	Houston TX	1-800-638-4266
Captree Chemical	W Babylon NY	1-800-899-2725
Carlton Chemical	Ft Lee NJ	1-800-772-1199
Carlyle Products Inc	Greenville SC	1-800-882-1520
Carolina PCO Supply	Columbus GA	1-800-233-8636
Carus Chemical Co	Peru IL	1-800-435-6856
Cedar Chemical	Memphis TN	1-800-423-8629
Cenex LOL Express Centers	Kennewick WA	1-800-535-3999
Cenex Land O Lakes Chemical Express Cen	Oslo MN	1-800-537-7340
Cenex Land O Lakes Express Center	Moorehead MN	1-800-633-5003
Certified Products	Jersey City NJ	1-800-654-2436
Champion Bill	Grenada MS	1-800-232-3594
Charter Chemical Inc	Atlanta GA	1-800-832-0839
Chas H Lilly	Portland OR	1-800-523-3668
Chem Central Dye Divison	Romulus MI	1-800-338-0684
Chem Genes Lab	Needham MA	1-800-762-9323
Chem Line Inc	Plano TX	1-800-443-2538
Chem Mart	Arlington TX	1-800-982-2436
Chem Pace	Toledo OH	1-800-423-5350
Chem Pro	Spartanburg SC	1-800-835-3712
Chem Sources Inc	Valencia CA	1-800-235-0705
Chem-Tech	Oceanside CA	1-800-437-0677
Chemco Intl	Greenville SC	1-800-253-0775
Chemcraft	Belmont MS	1-800-633-6281
Chemdal Corp	Palatine IL	1-800-243-6325
Chemdal	Palatine IL	1-800-992-5901
Chemical Abstracts Service		1-800-848-6538
Chemical Distributors	Westmont IL	1-800-533-3287
Chemical Monitoring Bureau	Round Lake IL	1-800-457-4280
Chemical Monitoring Bureau	Round Lake IL	1-800-523-6242
Chemical Products Co Inc	Omaha NE	1-800-445-9344
Chemical Products Group The	Philadelphia PA	1-800-523-5005
Chemical Sales Service	Worcester MA	1-800-878-2436
Chemical Solvents	Cleveland OH	1-800-362-0693
Chemical Specialties Corp	Silsbee TX	1-800-892-2436
Chemical Specialties Intl	Shingle Springs CA	1-800-541-1189
Chemical Ways	Arlington TN	1-800-222-7708
Chemicals & Solvents	Roanoke VA	1-800-523-3099
Chemland Inc	Des Plaines IL	1-800-238-7261
Chemlime NJ	Hodgkins IL	1-800-486-1338
Chemotive	Kennesaw GA	1-800-424-6248
Chemtech Industries Inc	Memphis TN	1-800-238-7433
Chemtech Industries Inc	Memphis TN	1-800-325-4864
Chemtech Industries Inc	Memphis TN	1-800-492-4981
Chemtech Industries Inc	Memphis TN	1-800-641-4214
Chidley & Peto Co The	Arlington Heights IL	1-800-662-3800
Chilean Nitrate Corp Industrial Orders	Norfolk VA	1-800-648-6827
Chlorinators & Controls	Escondido CA	1-800-245-6726
Chomerics-A W R Grace Co	Woburn MA	1-800-225-1936
Church & Dwight Co Inc	Princeton NJ	1-800-221-0453
Church & Dwight Co Inc	Princeton NJ	1-800-631-5591
Ciba Geigy Additives	Hawthorne NY	1-800-431-2360
Ciba Geigy Corps	Hawthorne NY	1-800-431-1900
Ciba Geigy P & A Divisions	Norcross GA	1-800-241-8811

Ciba Geigy Photo Poylmers Corp	Tarrytown NY	1-800-942-7829
Ciba Geigy	Houston TX	1-800-231-0646
Ciba Geigy	Oak Brook IL	1-800-323-7386
Citi Chem	Cherry Hill NJ	1-800-468-1883
Claremont Chemicals	Claremont NH	1-800-782-4541
Cleaning Solution Inc	Altamonte Springs FL	1-800-243-6911
Cleaning Systems Inc	Green Bay WI	1-800-225-2231
Clearwater Chemical	Clearwater FL	1-800-330-6121
Coastal Chemical	Pasadena TX	1-800-535-1561
Cole Growers	Dixon IL	1-800-435-9632
Coleman Oil Corp	Peoria IL	1-800-322-6145
Columbian Chemicals Co	Akron OH	1-800-336-2068
Columbian Chemicals	Atlanta GA	1-800-235-4003
Columbian Chemicals	St Louis MO	1-800-325-3604
Columbian Chemical		1-800-257-5076
Commerce Industrial Chemicals Inc	Milwaukee WI	1-800-242-7091
Commonwealth Chemical Inc	Louisville KY	1-800-438-3238
Computer Communications Of America	Troy MI	1-800-253-0756
Consolidated Chemical Industries	Houston TX	1-800-492-2990
Conspec Mktg & Mfg	Kansas City KS	1-800-348-7351
Continental Products Of Texas	Odessa TX	1-800-592-4684
Cooper Co Inc The	Gulf Breeze FL	1-800-634-4008
Copeq Trading	Houston TX	1-800-364-3330
Coral Intl	Waukegan IL	1-800-228-4646
Coretank Inc	Houston TX	1-800-348-5450
Cornbelt Chemical Co	Garden City KS	1-800-282-2254
Cornbelt Chemical Co	Garden City KS	1-800-545-3775
Cornbelt Chemical Co	Garden City KS	1-800-652-9306
Cornbelt Chemical	Powell WY	1-800-457-3491
Corroless North America Anti-Corrosion	Stamford CT	1-800-922-1991
Crest Industrial Chemicals Inc	Houston TX	1-800-833-8517
Crest Industrial Chemicals	Houston TX	1-800-622-9006
Crompton & Knowles	Charlotte NC	1-800-323-4383
Crompton & Knowles	Charlotte NC	1-800-432-6188
Crompton & Knowles	Charlotte NC	1-800-438-4122
Crucible Chemicals	Greenville SC	1-800-845-8873
Custom Chemicides	Fresno CA	1-800-227-6163
D S M Desotech Inc	Elgin IL	1-800-222-7189
DRC	Denver CO	1-800-321-8824
Danville Express	Danville KS	1-800-662-4212
Datachem Salt Lake City Chemistry Labor	Murray UT	1-800-356-9135
Degussa Corp Chemical Catalyst Division	Ridgefield Park NJ	1-800-422-8773
Deltech Corp	Baton Rouge LA	1-800-535-9945
Detergent Mktg System	Aurora IL	1-800-227-7765
Dewolf Mktg Inc	E Providence RI	1-800-521-0065
Diagnostic Chemical Limited	Oxford CT	1-800-325-2436
Diversey Corp Southern Region	Houston TX	1-800-732-2446
Diversey Corp	Livonia MI	1-800-521-8140
Diversey Wyandotte Corp	Livonia MI	1-800-482-8010
Dober Chemical	Midlothian IL	1-800-323-4983
Dolphin Chemical & Supply	Houston TX	1-800-782-1718
Dominion Chemical	Petersburg VA	1-800-852-6970
Donald Riggsby	Mohomet IL	1-800-952-6294
Dover Chemical	Dover OH	1-800-321-8805
Dowelanco Employee Benefits	Indianapolis IN	1-800-447-9747
Duochem	Wheeling IL	1-800-222-0477
Dupage Sales & Service	Glen Ellyn IL	1-800-462-4801
Dupont Customer Information Center	Wilmington DE	1-800-441-7515

Dymon Inc	Kansas City KS	1-800-255-4564
E M Diagnostic Systems	Gibbstown NJ	1-800-257-5541
EKA Nobel Inc	Columbus MS	1-800-821-9486
EM Diagnostics	Gibbstown NJ	1-800-443-3637
EM Science Division Of EM Ind	Gibbstown NJ	1-800-222-0342
EM Science	Cherry Hill NJ	1-800-922-1084
EXSL/Ultra Labs	Hayward CA	1-800-545-4536
Eaglebrook Inc	Lakeshore MS	1-800-525-7109
Earnest Machine Products Co	Tampa FL	1-800-237-1671
Earnest Machine Products Co	Tampa FL	1-800-631-5275
East Bay Oil	Hayward CA	1-800-426-9097
Ecochem Co	Adell WI	1-800-252-2842
Ecolab, Inc.	St Paul MN	1-800-352-5326
Econo-Chem Inc	Metairie LA	1-800-421-0568
Electro Chemicals	Maple Plain MN	1-800-321-9050
Elevation Corp	Astoria NY	1-800-368-7909
Elf Atochem North America Inc	Fresno CA	1-800-331-0252
Elf Atochem North America Inc/Decco	Philadelphia PA	1-800-221-0925
Elf Atochem North America Inc	Crosby TX	1-800-526-5544
Elf Atochem North America	Philadelphia PA	1-800-328-2811
Elf Atochem North America	Philadelphia PA	1-800-932-0420
Emerald City Chemical	Seattle WA	1-800-431-5511
Emergency Chemical Inc	Orlando FL	1-800-548-4571
Emerson & Cuming Inc	Canton MA	1-800-225-9936
Emery Chemicals	Cincinnati OH	1-800-543-7370
Emulsion Engineering	Sanford FL	1-800-327-6270
Energy & Environmental Service Limited	Oklahoma City OK	1-800-635-7716
Entech Recovery Inc	Union City CA	1-800-352-5217
Enterprise Chemical	Mchenry IL	1-800-262-2329
Enterprise	Freeport TX	1-800-231-3384
Enthone-OMI Inc	Warren MI	1-800-521-4267
Enviro Care Chemicals	W Bend WI	1-800-641-6981
Environmentally Sensitive Products Inc	Atlanta GA	1-800-532-3513
Envirosafe	Redwood City CA	1-800-227-9744
Expert Products	Moorpark CA	1-800-225-6929
Extractor Chemicals	Muskegon MI	1-800-338-2436
F & C Division Douglas & Lomason Co	Carrollton GA	1-800-241-4051
F M C Lithium Division	Gastonia NC	1-800-362-2549
F M C Material Handling Equip Division	Downers Grove IL	1-800-323-4834
FIUKA Chemical	Ronkonkoma NY	1-800-358-5287
FMC AG Chemical Group	Omaha NE	1-800-231-5808
FMC AG Chemical Group	Omaha NE	1-800-346-0833
FMC Corp Chemical Products Group	Pleasanton CA	1-800-538-3968
FMC Corp Marine Colloids Division	Philadelphia PA	1-800-346-5101
FMC Food & Pharmaceutical Products Divi	Philadelphia PA	1-800-362-3773
FMC Food Grade Phosphate Technical	Princeton NJ	1-800-848-3362
FMC	Charlotte NC	1-800-432-6858
FMC	Charlotte NC	1-800-438-1854
Fairfield Chemical Co	Blythewood SC	1-800-845-2715
Fang Chemicals	Carrollton GA	1-800-582-5169
Farmco Industry	San Gabriel CA	1-800-423-3396
Fibre Pro	Colorado Springs CO	1-800-525-3896
Filtrol Corp	Los Angeles CA	1-800-421-1889
Fine Line Chemicals	Glendale WI	1-800-962-8982
Fisher Stevens Inc	Victoria TX	1-800-448-3760
Force Chemicals Industries	Paoli PA	1-800-647-3575
Foresight Intl Specialty Chemicals	Omaha NE	1-800-637-1344
Foseco	Chehalis WA	1-800-426-3489

Four Corners Weed Control	Farmington NM	1-800-354-6716
Four Star Chemical	Los Angeles CA	1-800-243-6264
Franklin Miller Inc	Livingston NJ	1-800-932-0599
French Color & Chemical Co	Englewood NJ	1-800-762-9098
Fuller Sales	Dalton GA	1-800-628-1661
GEQ	Fenton MI	1-800-336-5869
Gala Chemical	Phoenix City AL	1-800-842-4585
Gallade Chemicals Inc	Santa Ana CA	1-800-451-4661
Gantrade Corp	Montvale NJ	1-800-426-8723
Gas Treating Chemicals	Nadine NM	1-800-262-8263
Gator Supply & Equip	Tampa FL	1-800-330-5256
Gecor Industries Inc	Grayson GA	1-800-552-0883
Gemini Mfg	Hollywood FL	1-800-342-1200
Genencor Intl Inc	Rochester NY	1-800-847-5311
General Chemical Orchard Park Ny	Pittsburg CA	1-800-247-4519
General Chemical	Syracuse NY	1-800-255-7589
General Chemical	Parsippany NJ	1-800-631-8050
Genii USA	Apple Valley MN	1-800-332-5518
George Mann & Co	Providence RI	1-800-556-2426
Georgia Auto Magic	Augusta GA	1-800-932-4643
Georgia Gulf Inc	Plaquemine LA	1-800-822-7529
Glotex Chemicals	Roebuck SC	1-800-277-1005
Goldkist	Alcolu SC	1-800-824-3994
Great Lakes Chemical	W Lafayette IN	1-800-428-7947
Great Western Chemical Co		1-800-683-9724
Great Western Chemical Co		1-800-942-0580
Great Western Chemical		1-800-544-2436
Great Western Chemical		1-800-852-5409
Green Chemical	Winchester VA	1-800-368-3713
Green Valley Chemicals	Portsmouth OH	1-800-767-5057
Greenway Chemical Co Inc	Knoxville TN	1-800-258-5829
Griffin Corp	Houston TX	1-800-621-2390
Griffin Customer Fax	Valdosta GA	1-800-446-8655
Griffin	Valdosta GA	1-800-237-1854
HCI Chemicals USA	Houston TX	1-800-231-7818
HCI Chemtech St Louis Branch	St Louis MO	1-800-525-1912
Halliburton Industrial Service	Bossier City LA	1-800-932-5236
Halls Chemical Inc	Tunnel Hill GA	1-800-537-8664
Halsen Products	Belmont MS	1-800-344-6696
Hanlin Group Inc	Linden NJ	1-800-526-7616
Harshaw Filtrol	Solon OH	1-800-424-2015
Harshaw Filtrol	Pepper Pike OH	1-800-752-8464
Hartford Chemical	Westbury NY	1-800-992-2729
Harwick Chemical Corp	Norcross GA	1-800-824-4063
Hawk Creek Laboratory Inc/ Reagent Chem	Glen Rock PA	1-800-637-2436
Haze Line Intl	Coral Springs FL	1-800-238-1146
Heartland Ag	Vandalia IL	1-800-230-0488
Heico Chemicals Inc	Del Water Gap PA	1-800-344-3426
Helena Chemical Co	Watervliet MI	1-800-457-7018
Helena Chemical	Dickinson ND	1-800-227-8389
Helena Chemical	Phoenix AZ	1-800-238-5992
Helena Chemical	Phoenix AZ	1-800-338-8260
Helena Chemical	Phoenix AZ	1-800-422-0872
Helena Chemical	Saginaw MI	1-800-543-6262
Helena Chemical	Union Gap WA	1-800-648-8836
Helena Chemical	Bluffton IN	1-800-972-7866
Henkel Corp Hdqrs Philadelphia	King Of Prussia PA	1-800-521-5317
Henkel Corp Process Industries	Ambler PA	1-800-322-0919

Henkel Corp	Bellevue WA	1-800-443-6611
Hercules Inc	Wilmington DE	1-800-445-9490
Hercules Inc	Middletown NY	1-800-828-3366
Hercules Product Information	Wilmington DE	1-800-247-4372
Hercules Stockholder Services	Wilmington DE	1-800-441-9029
Heyltex	Houston TX	1-800-237-6793
Hi Strand Chemical	Lenoir NC	1-800-523-8136
High Point Chemical Technical Service	High Point NC	1-800-477-2214
High Point Chemical	High Point NC	1-800-727-2214
High Purity Chemical	Meridian ID	1-800-232-4722
Highside Chemicals	Gulfport MS	1-800-359-5599
Hilord Chemical	Hauppauge NY	1-800-645-1022
Hoechst Celanese Product Information Ce	Edison NJ	1-800-235-2637
Holchem	Fresno CA	1-800-333-7965
Holtrachem Inc	Natick MA	1-800-343-6470
Holtrachem Inc	Natick MA	1-800-343-8430
Holtrachem West	Anaheim CA	1-800-342-2436
Hopton Technologies	Rome GA	1-800-231-7702
Hub States	Indianapolis IN	1-800-428-4416
Hubbard Hall	Inman SC	1-800-442-5573
Huge Co Inc	St Louis MO	1-800-325-3371
Huls America Plastics Division	Somerset NJ	1-800-344-6080
Humphrey Chemical Co The	North Haven CT	1-800-652-3456
Huntington Lab Inc	Huntington ID	1-800-448-6522
Huntsman Chemical Corp Baton Rouge	Baton Rouge LA	1-800-525-2205
Huntsman Chemical	Chesapeake VA	1-800-237-8653
Hydro Systems	Cincinnati OH	1-800-543-7184
Hydro-Balance	Carrollton TX	1-800-527-5166
IBM Chemical Control Information	New York NY	1-800-426-4333
ISK Biotech Corp	Marietta GA	1-800-525-1500
ISK Biotech	Fresno CA	1-800-525-4485
Ideal Chemical & Supply	Memphis TN	1-800-232-6776
Ideal Chemical & Supply	Memphis TN	1-800-824-0356
Ideas	Chicago IL	1-800-274-2010
Illini SS Inc	Tolono IL	1-800-328-6810
Infotrac Chemical Emergency Response Sy	Eden Prairie MN	1-800-326-5640
Inland Technology	Tacoma WA	1-800-552-3100
Inolex Chemical Co	Addy WA	1-800-521-9892
Integrated Separation Systems	Natick MA	1-800-433-6433
Intercat	Manasquan NJ	1-800-346-5425
Intercon Chemical Co	St Louis MO	1-800-325-9218
Intergen	Purchase NY	1-800-468-7436
International Chemical	Memphis TN	1-800-238-6531
International Speciality Chemicals Inc	Mt Olive NJ	1-800-453-0397
Interox America	Houston TX	1-800-468-3769
Interpolymer	Canton MA	1-800-262-1281
Interstate Chemical Co Inc	Pueblo West CO	1-800-422-4485
Interstate Research	Sterling Heights MI	1-800-247-8055
Intex Products	Greenville SC	1-800-845-1668
Isotec Specialty Chemicals	Miamisburg OH	1-800-448-9760
JBH Mfg	Corsicana TX	1-800-446-2091
JLB Intl Chemical	Vero Beach FL	1-800-228-1833
JMN Specialties	Belle Chasse LA	1-800-547-6344
Jacam Chemical	Sterling KS	1-800-248-0357
Joebear Inc	Toledo OH	1-800-445-2720
Jungbunzlauer Inc	Newton MA	1-800-828-0062
Kalama Chemical Inc	Kalama WA	1-800-233-7799
Kalama Chemical Inc	Kalama WA	1-800-742-6147

Keil Chemical	Hammond IN	1-800-628-9079
Kemira Inc	Savannah GA	1-800-453-6472
Kenflo Products Inc	Milledgeville GA	1-800-972-2414
King E & F & Co Inc	Norwood MA	1-800-532-6003
King Industries	Norwalk CT	1-800-423-0524
Kleenrite Chemical	Carson CA	1-800-553-3678
Koch Sulfur Products	Bettendorf IA	1-800-282-6819
Koppers Indutries Inc/Sales	Pittsburgh PA	1-800-321-9876
Kraft Chemical Co	Melrose Park IL	1-800-345-5200
La-Mar-Ka Chemical	Baton Rouge LA	1-800-826-5959
LaRoche Industries	Columbus GA	1-800-654-7930
Lamotte Chemicals	Chestertown MD	1-800-344-3100
Lancaster Synthesis Ltd	Pelham NH	1-800-238-2324
Land & Sea Labs	Rockport TX	1-800-420-8345
Land & Sea Products	Grand Rapids MI	1-800-321-1548
Lang's Standard Solutions Inc	Baton Rouge LA	1-800-452-4563
Laroche Chemicals	Baton Rouge LA	1-800-458-7404
Laroche Chemicals	Baton Rouge LA	1-800-528-4963
Lawrence & Associates	Hickory NC	1-800-542-1521
Lebanon Chemical Co	Danville IL	1-800-637-2101
Lebanon Chemical	Avon Heights PA	1-800-233-0628
Lebanon Chemical	Avon Heights PA	1-800-637-5190
Liqui-Tech Corp	Coal City WV	1-800-637-4527
Loctite Technical Info	Newington CT	1-800-562-8483
Lonza	Fair Lawn NJ	1-800-631-3647
Lubar Chemical Co/Janitorial Chemicals	Kansas City MO	1-800-821-5207
Lubrizol Corp California Sales Office	Whittier CA	1-800-438-6953
MCT Products	Gainesville GA	1-800-526-4288
MDM Scientific	Stafford TX	1-800-245-7549
MVR Chemical	Skillman NJ	1-800-833-8342
Madison Bionics	Tempe AZ	1-800-526-3051
Mangren Research & Development Corp	Mansville TX	1-800-433-5379
Mantek Chemicals	Irving TX	1-800-527-7850
Marinize Products	N Miami FL	1-800-842-4380
Marlowe Van Loan	High Point NC	1-800-422-4685
Marsulex Inc	Burr Ridge IL	1-800-446-0182
Martex Industrial Chemicals	Prior Lake MN	1-800-662-1986
Marx Lab	Louisville KY	1-800-762-4316
Maryland Chemical Co Inc	Baltimore MD	1-800-292-1967
Mason Chemical	Arlington Heights IL	1-800-362-1855
Master Builders Inc	Cincinnati OH	1-800-458-3921
Master Builders	Cleveland OH	1-800-227-3350
Mathieu Hampden	Schenectady NY	1-800-447-7471
May Day Chemical	Kalamazoo MI	1-800-446-6525
Mayo Chemical Co Inc	Smyrna GA	1-800-962-6296
Mazer Chemicals	Gurnee IL	1-800-323-0856
Mega Power	Martinez CA	1-800-624-3002
Mercury Chemical	Burlington NC	1-800-637-2461
Metro Chem	South Kearny NJ	1-800-332-7627
Metsa Serla Chemicals	Morrow GA	1-800-638-7224
Micro Care Chemical Corp	Bristol CT	1-800-638-0125
Micro Flo	Lakeland FL	1-800-543-3120
Micron Diagnostics	Fairfax VA	1-800-642-7660
Mid Co Chemical Co	Bay City TX	1-800-559-6432
MidAmerica Research & Chemical	Columbbus NE	1-800-228-8508
Midlantic Biomedical	Paulsboro NJ	1-800-441-0366
Midwest Biotech	Fishers IN	1-800-772-2798
Mike Chemical	Newland NC	1-800-648-6453

Mil Tex	Salt Lake City UT	1-800-828-8321
Miles, Inc.	Pittsburgh PA	1-800-348-7414
Miles, Inc.	Pittsburgh PA	1-800-422-1559
Miles, Inc.	Pittsburgh PA	1-800-638-6061
Miles, Inc.	Pittsburgh PA	1-800-882-9987
Milwaukee Solvents & Chemical	Milwaukee WI	1-800-352-3560
Milwaukee Solvents & Chemical	Des Moines IA	1-800-798-6019
Mooney Chemicals	Cleveland OH	1-800-321-9696
Morristown Chemicals Group Inc	Flanders NJ	1-800-822-1441
Moss Soap And Chemical Company	Miami FL	1-800-752-4654
Motsendocker Advanced Development	San Diego CA	1-800-346-1633
Mount Hood Chemical Corp	Portland OR	1-800-547-2594
Murrell's Whsle And Mft Inc	Grayson GA	1-800-428-8740
NAIAD	Pleasanton CA	1-800-541-6662
NCC	Marietta GA	1-800-241-4547
Nalfleet Bull and Roberts Inc	Springfield NJ	1-800-535-1896
Narchem	Chicago IL	1-800-458-1057
National American Sales	Thibodaux LA	1-800-821-3732
National Magnesia Chemicals	Moss Landing CA	1-800-622-2499
National Specialty Gases	Durham NC	1-800-346-8238
Nationwide Cleaning Supplies Inc	Sioux Falls SD	1-800-356-7320
Nationwide Cleaning Supplies	Sioux Falls SD	1-800-468-2580
New England Reagent Lab	E Providence RI	1-800-556-7575
Nijacol Products Inc	Ashland MA	1-800-438-7657
Nitram	Tampa FL	1-800-237-6956
Non Haz Alternatives	Wooster OH	1-800-331-3688
Norman Fox & Co	Los Angeles CA	1-800-632-1777
North American Chemical	Overland Park KS	1-800-637-2775
North American Research Corp	Carrollton TX	1-800-527-7520
North Chemical	Marietta GA	1-800-241-2367
Novacor Chemicals Inc	Parsippany NJ	1-800-831-2244
Novamay Technologies	Livonia MI	1-800-521-5770
Novick Chemical Co	Clark NJ	1-800-982-5584
Nu-Calgon Whsle Inc	St Louis MO	1-800-554-5499
Nufarm	St Joseph MO	1-800-852-5234
Nutech Services	Counce TN	1-800-962-1901
Oakite Products Inc	Houston TX	1-800-231-5040
Oakite Products	Industry CA	1-800-221-9467
Oakite Products	Industry CA	1-800-352-8985
Oakite Products	Romulus MI	1-800-521-6200
Oakite	Berkeley Heights NJ	1-800-526-4473
Occidental Chemical Corporation	Dallas TX	1-800-828-1144
Oilfield Treating Chemicals Inc	Raymondville TX	1-800-882-4178
Olin Hunt Specialty Inc	W Paterson NJ	1-800-553-6546
P M C	Rocky River OH	1-800-543-2466
PBI Gordon Corp	Kansas City MO	1-800-821-7925
PCR Inc	Gainesville FL	1-800-331-6313
PDM Inc	Wilmington DE	1-800-288-0768
PMC Specialties Group	Rocky River OH	1-800-227-2442
PMC Specialties Group	Rocky River OH	1-800-228-3673
PQ Corp	Utica IL	1-800-338-6109
Par Antifreeze	Westmont IL	1-800-325-7793
Parker Amchem	Horsham PA	1-800-235-2172
Parker Division HCPC	Troy MI	1-800-521-1355
Pavco	Cleveland OH	1-800-321-7735
Peach State Labs	Rome GA	1-800-634-1653
Pennwalt	Cornwells Heights PA	1-800-523-2762
Pennzoil Sulphur	Pecos TX	1-800-662-2048

Penreco	Dickinson TX	1-800-458-5845
Peridot Chemicals Inc	Augusta GA	1-800-545-2586
Pharmco Products Inc	Brookfield CT	1-800-243-5360
Phillips Specialty Chemical Co	Borger TX	1-800-858-4327
Phillips William E	Glen Ellyn IL	1-800-331-2940
Photographers Formulary	Condon MT	1-800-922-5255
Pico Chemical Corp	Chicago Heights IL	1-800-472-0669
Pioneer Chlor Alkali	Henderson NV	1-800-334-9503
Plasma Chem Associate	Bradley Beach NJ	1-800-343-0437
Plasmine Technology	Pensacola FL	1-800-752-6225
Ply Bond Chemical	Virginia BeachVA	1-800-462-0358
Pneumateck Industries	Bethlehem PA	1-800-553-6307
Polychrome Chemical Cellomer Division	Newark NJ	1-800-548-5456
Polysciences INC	Warrington PA	1-800-523-2575
Polyurethane Specialties Inc	Lyndhurst NJ	1-800-348-2553
Prez Chem Inc	Columbia SC	1-800-433-1999
Pride Solvents And Chemical Co	W Babylon NY	1-800-645-5506
Products Chemical	Cleveland OH	1-800-233-5078
Promaster Division Bishop Carpet Care	Dothan AL	1-800-231-4461
Pronova Biopolymer Inc	Portsmouth NH	1-800-223-9030
Puget Sound Specialty Products	Tacoma WA	1-800-422-0157
Q O Chemicals	W Lafayette IN	1-800-621-9521
Quaker Chemical Corp.	Conshohocken PA	1-800-458-2436
Quality Controlled Biochemi Cals Inc	Hopkinton MA	1-800-435-2080
Quantum Chemical	Cincinnati OH	1-800-543-5900
RML Corp	Waukesha WI	1-800-542-2750
Rainbow Mfg	Pelham AL	1-800-637-6047
Raisio Inc Paper Chemical Manufacturer	Berwick PA	1-800-847-2812
Raisio Inc Paper Chemical Manufacturer	Berwick PA	1-800-847-6772
Ramco Chemicals Inc	W Bridgewater MA	1-800-448-4861
Rauenzahn Raymond	Horth Hills PA	1-800-626-9369
Reagent Chemical Operation & TransportaSt	Gabriel LA	1-800-535-9985
Reagent Chemical	Houston TX	1-800-322-7855
Reagent Chemical	Stanton TX	1-800-343-3710
Red Bird Service	Osgood IN	1-800-428-3502
Reef Chemical	Snyder TX	1-800-552-0691
Regency Chemical Corp	Holiday FL	1-800-922-9055
Regis Chemical	Morton Grove IL	1-800-323-8144
Reichhold Chemicals Inc Coating Polymer	Research Triangle Park NC	1-800-292-7888
Reichhold Chemicals Inc	Research Triangle Park NC	1-800-346-1165
Reichhold Chemical	Research Triangle Park NC	1-800-544-1870
Reichold Chemical Inc Order Dept	Research Triangle Park NC	1-800-874-5403
Reicholds Chemicals Inc	Durham NC	1-800-431-1920
Reilly Whiteman	Conshohocken PA	1-800-533-4514
Reilly-Whiteman Southern Division	Greensboro NC	1-800-343-3047
Release Coating Of New York	Wellsville NY	1-800-457-7817
Remove It Products	Houston TX	1-800-645-0833
Resinall	Severn NC	1-800-421-0561
Rhone Poulenc AG Co Product Use Informa	Beltsville MD	1-800-334-9745
Rhone-Poulenc Agricultural Products	Research Triangle Park NC	1-800-334-7577
Rockland React-Rite	Rockmart GA	1-800-221-4799
Rockville Chemical Corp	Farmingdale NY	1-800-342-6716
Rohm & Haas Illinois Inc	Illipolis IL	1-800-843-2820
Rohme & Haas Texas	Deer Park TX	1-800-782-8579
S & S Chemical Corp	Calumet Park IL	1-800-621-5319
S U S Chem-Mark Corp	Rock Hill SC	1-800-343-7872
SCM Chemicals	Baltimore MD	1-800-638-3234
SKW Alloys	Niagara Falls NY	1-800-828-6621

SQM Iodine Corp USA	Norfolk VA	1-800-532-8213
Sacco Pigments	Forest Park IL	1-800-826-3018
Saf Bulk Sales	Ronkonkoma NY	1-800-321-0723
San Esters Corp-Monomers	New York NY	1-800-337-8377
Sandoz Chemicals Corporation	Charlotte NC	1-800-631-8077
Sandoz Crop Protection Inc	Des Plaines IL	1-800-553-4833
Sani Kleen Chemicals	Miami FL	1-800-654-4624
Scholle Corp	College Park GA	1-800-535-2968
Schoofs Inc	Moraga CA	1-800-227-1345
Schoonover	Atlanta GA	1-800-331-2808
Schulze Processing	Shreveport LA	1-800-592-6940
Schweizerhall	Piscataway NJ	1-800-243-6564
Scientific Intl Research Min-Kem	Maple Lake MN	1-800-445-1885
Scotch Corp	Dallas TX	1-800-334-2077
Seeler Industries Inc	Joliet IL	1-800-336-2422
Seidler Chemical	Newark NJ	1-800-848-8192
Sentry Polymers	Freeport TX	1-800-231-2544
Sequa Chemicals Order Department	Chester SC	1-800-428-7855
Shipley Co Inc	Carrollton TX	1-800-527-3730
Shipley Co	Newport Beach CA	1-800-854-7545
Sierra Chemical	Battle Mountain NV	1-800-777-2495
Sierra Chemical	Battle Mountain NV	1-800-777-8965
Sierra Chemical	Urbana OH	1-800-782-9339
Sigma Chemical - Clinical Technical Ser	St Louis MO	1-800-325-0250
Sigma Chemical - Clinical Technical Ser	St Louis MO	1-800-325-3010
Silenus Lab Limited	Miami FL	1-800-932-4107
Simmons Chemical Corp	Sarasota FL	1-800-426-3265
Sipsy Chemical	Trenton NJ	1-800-327-4779
Smith Chemical	Canton OH	1-800-255-1106
Smith Chemical	Canton OH	1-800-433-6191
Solkatronic Chemicals	Fairfield NJ	1-800-521-3981
Solkatronic Chemicals	Los Gatos CA	1-800-774-7234
Southeastern Road Treatment	Evans GA	1-800-447-3781
Southern Chemical Inc	Hattiesburg MS	1-800-247-7793
Southland Corp	Summit IL	1-800-323-3231
Southwest Petro Chem Inc	Olathe KS	1-800-527-2529
Sovereign Chemical Co	Cuyahoga Falls OH	1-800-831-1722
Standard Adhesive & Chemical Co Inc	Dalton GA	1-800-252-4851
Standard T	Chicago Heights IL	1-800-533-3438
Stat Enterprises Division Of Pride Lab	Farmingdale NY	1-800-645-9198
Stepan Co	Winder GA	1-800-228-8312
Stepan Co	Northfield IL	1-800-457-7673
Stepan Co	Winder GA	1-800-523-3614
Stephenson Chemical Co Inc	College Park GA	1-800-241-3343
Stockhausen, Inc.	Greensboro NC	1-800-334-0242
Stoner	Quarryville PA	1-800-227-5538
Strem Chemical	Newburyport MA	1-800-647-8736
Suffolk Sales & Service	Suffolk VA	1-800-628-5185
Sulphuric Acid Trading Co	Tampa FL	1-800-633-1358
Summit Chemical Co	Baltimore MD	1-800-227-8664
Summit Specialty Chemical Corp	Charlotte NC	1-800-645-4529
Sunapee Chemical	Wooster OH	1-800-262-0290
Sunette Center	Edison NJ	1-800-344-5807
Suntech Inc	Castle Rock CO	1-800-682-4790
Supelco	Bellefonte PA	1-800-247-6628
Superior By Pas & Supply	Alpine MI	1-800-323-0344
Superior Evaporant	San Jose CA	1-800-927-8737
Supertech Products Inc	Ft Lauderdale FL	1-800-245-0468

Synergy Fluids	Houston TX	1-800-533-8062
Syntex Agribusiness	Springfield MO	1-800-346-8170
Synthetic Products Co	Cleveland OH	1-800-321-4236
Synthetic Products Co	Cleveland OH	1-800-421-8813
Synthetics Products Co	Stratford CT	1-800-445-4924
T&R Chemicals Inc	Clint TX	1-800-351-6025
TAB Chemicals	Chicago IL	1-800-331-6824
TS Distributor	Darien CT	1-800-499-0150
Talco Industrial Chemicals	Ooltewa TN	1-800-648-5549
Target Specialty Products	Phoenix AZ	1-800-352-5548
Tastemaker	Cincinnati OH	1-800-892-1199
Tech Chem	Stockton MO	1-800-852-5641
Technichem	Boise ID	1-800-635-8930
Technology Chemical Inc Southern Califo	Oakland CA	1-800-548-0770
Technology Chemical Inc Southern Califo	Oakland CA	1-800-845-0770
Tenneco Minerals	Green River WY	1-800-443-2785
Terra International	Auburn IL	1-800-343-2617
Terra Intl Inc	Cedarville OH	1-800-468-2436
Terra Intl	Fayette MO	1-800-247-7974
Terra Intl	Morganfield KY	1-800-542-3244
Terra Intl	York PA	1-800-762-3837
Texas Aniline Dye	Hempstead TX	1-800-346-5352
Textile Chemical	Reading PA	1-800-422-8426
Theochem Lab	Tampa FL	1-800-237-2591
Therminol Heat Transfer Fluids	St Louis MO	1-800-433-6997
Thiotech USA	Houston TX	1-800-633-6197
Timerland Enterprises	Monticello AR	1-800-752-7009
Tip Top Distribution	Goose Creek SC	1-800-782-4879
Tip Top Teak	Ripley MS	1-800-847-8671
Titan Chemical	Houston TX	1-800-648-8801
Tomco Chemical	Wantagh NY	1-800-645-3285
Total Systems	Las Vegas NV	1-800-323-2313
Traverse Chemicals Intl	LANSING MI	1-800-247-6136
Trinity Mfg Inc	Hamlet NC	1-800-632-6228
Trinity Solutions	Peachtree City GA	1-800-368-1201
Tuff Coat	Loganville GA	1-800-578-8555
U O P	Des Plaines IL	1-800-243-6757
UAP/GA Ag Chem Inc	La Grange NC	1-800-832-0044
UCB Chemical Corp	Norfolk VA	1-800-426-3820
UCR	Indianapolis IN	1-800-382-9783
US Movidyn Corp	Chicago IL	1-800-621-4026
Ultra Additives	Paterson NJ	1-800-524-0055
Uni-Chem	Ft Lauderdale FL	1-800-448-8642
Unichem Corp	Chicago IL	1-800-862-2436
Union Camp	Jacksonville FL	1-800-874-9220
Union Carbide Specialty Chemicals Div-R	Marietta GA	1-800-525-6648
Unique Chemical & Waterworks Products C	Schulenburg TX	1-800-637-2279
Uniroyal Chemical Co Inc	Farmers Branch TX	1-800-527-3736
Uniroyal Chemical Co Inc	Middlebury CT	1-800-962-4260
Uniroyal Chemical Specialty Chemical	Middlebury CT	1-800-322-3243
Unisol Chemicals	Asheboro NC	1-800-682-2120
United Horticultural Supply	Columbus OH	1-800-222-9963
United Lab	St Charles IL	1-800-323-2594
United Lab	St Charles IL	1-800-323-7137
Universal Scientific Of Arizona	Mesa AZ	1-800-253-5091
Universal Scientific Supply	Greensboro NC	1-800-828-8648
VI-Chem	Medley FL	1-800-325-6109
Varichem Inc Division E A Products	Bay City TX	1-800-833-5394

Verner Chemical Corp	Richardson TX	1-800-426-7713
Versatile Products	Kosciusko MS	1-800-942-4121
Vibrant Products	Marietta GA	1-800-972-0175
Vista Products	Indianapolis IN	1-800-452-0035
W. A. Cleary Chemical	Somerset NJ	1-800-524-1662
WI-Chem	Chattanooga TN	1-800-448-7767
WIM Inc	Cleveland OH	1-800-772-3048
Wacker Silicones Corp	Oakbrook Terrace IL	1-800-521-4566
Wacker Silicones	Charlotte NC	1-800-521-3698
Wacker Silicones	Irvine CA	1-800-541-9517
Web-Away Inc	Ocala FL	1-800-562-3415
Webb	Newborn GA	1-800-342-9322
Wechem Inc	Jefferson LA	1-800-426-0512
Wel-Flo Chemical	Pettus TX	1-800-245-9203
Wepak Corp	Charlotte NC	1-800-438-4270
Western Reserve Chemical	Stow OH	1-800-321-2676
Western Technology	Boise ID	1-800-235-7107
Westrand Chemicals	Sparks NV	1-800-544-1852
Weststar Five Inc	Boise ID	1-800-524-1750
White All Lab	Wylie TX	1-800-448-7882
White Chemical Intl	Stafford TX	1-800-251-2299
Whittaker Clark & Daniels Inc	S Plainfield NJ	1-800-732-0562
Whittaker Clark & Daniels Inc	Plainfield IL	1-800-833-8140
Whittaker Clark & Daniels Inc	Norcross GA	1-800-833-8142
Whittaker Clark & Daniels Inc	Laguna Hills CA	1-800-843-1935
Wilbur-Ellis Co Southern Division	Eagle Lake TX	1-800-255-1948
Windsor Wax	Hoboken NJ	1-800-243-8929
Witco Corp	Jefferson LA	1-800-833-7686
Witco	Greenwich CT	1-800-759-4826
Witco	Greenwich CT	1-800-850-3135
Wright Chemical	Columbus GA	1-800-508-9497
YMC Inc	Wilmington NC	1-800-962-6311
Yorkshire Nachem Inc	Rockland MA	1-800-622-4361
Yorkshire Nachem	Peabody MA	1-800-220-7664
Zeeland Chemicals Inc	Zeeland MI	1-800-223-0453
Zeneca Resins	Wilmington MA	1-800-458-0014
Zeochem	Spring TX	1-800-237-9963
Zeon Chemicals USA Inc	Louisville KY	1-800-776-2460
Zircon	Cleveland OH	1-800-547-4328
Zobrist J C	Charleston WV	1-800-843-2990
Zymed Labs	S San Francisco CA	1-800-874-4494

CHEMISTS - ANALYTICAL & CONSULTANTS

Bart-Lind Corp The	Tukwila WA	1-800-553-5290
Biotech Research Institute	Congers NY	1-800-437-4246

CHEMICALS - MANUFACTURING

Alcoa Arkansas Chemical Sales	Bauxite AR	1-800-643-8771
Alpha Intermediates	Elmendorf TX	1-800-334-2480
Ausimont USA Inc	Thorofare NJ	1-800-323-2874
Cambridge Isotope Labs	Andover MA	1-800-476-8673
Camco Chemical Co.	Florence KY	1-800-354-1001
Catawba Charlab	Charlotte NC	1-800-763-6107
Chemtech Products Inc	St Louis MO	1-800-325-3332
Chemtex	Charlotte NC	1-800-532-5361
Ciba Additives	Troy MI	1-800-982-9340

Coral Intl	Paramount CA	1-800-899-6364
D C C Inc	Dallas TX	1-800-383-0507
Delta Omega Technologies	New Iberia LA	1-800-717-9922
Discovery Aluminas	Port Allen LA	1-800-335-5515
Elf Atochem North America	Philadelphia PA	1-800-225-7788
Esco Intl	Covington LA	1-800-831-3656
Essential Industries	Merton WI	1-800-551-9679
FMC ACG Specialty Products Order Center	Princeton NJ	1-800-321-3621
Fuller H B Automotive Products	Madison Heights MI	1-800-633-7789
I T & Q Chemical	Greenville SC	1-800-453-9218
ISK Biotech	Marietta GA	1-800-241-4128
Laurel Industries Inc	Cleveland OH	1-800-221-1304
Martin Marietta Magnesia Specialties In	Baltimore MD	1-800-648-7400
Oakwood Products Inc	W Columbia SC	1-800-467-3386
Peninsula Laboratories Inc.	Belmont CA	1-800-922-1516
Plainsman Technology	Marlow OK	1-800-256-4703
Silbond	Weston MI	1-800-543-3014
Snadoz	Charlotte NC	1-800-333-3635
Spectra Sol Inc	Warwick NY	1-800-933-9133
Synchem Inc	Aurora OH	1-800-882-9267
Ultrashield	Lauderhill FL	1-800-874-0841
Zeon Chemicals Inc	Rolling Meadows IL	1-800-735-3388

CHEMICAL MILLING
| F & H Chemicals | Tulare CA | 1-800-824-0743 |
| Image Technology | Placentia CA | 1-800-554-6243 |

CHEMICAL PLANT EQUIPMENT & SUPPLIES
Chem-Pro Inc	Evansville IN	1-800-892-4186
ChemIndustrial Systems Inc	Cedarburg WI	1-800-375-8551
Cordova Chemical Used Equip	Muskegon MI	1-800-422-3577
J Little Mercer Co. Inc	Rehoboth MA	1-800-637-2371
Steri Technologies	Bohemia NY	1-800-253-7140
Van Zyverdens Equip Sale	Willow Grove PA	1-800-942-9489
Zeochem	Glen Ellyn IL	1-800-248-6794

CHEMICALS - RECLAIMING
| JB Crawford Chemical | Crawfordsville AR | 1-800-822-5949 |
| M & J Solvents | Atlanta GA | 1-800-762-4098 |

ECO-PRENEURS - PRODUCTS & SERVICES:

Anti-Vivisection
National Anti-Vivisection Society — 1-800-888NAVS

Aromatherapy
Aromatherapy Products — 1800AROMATIC

Art
National Wildlife Galleries — 1-800-382-5278

Arts Supplies - Graphics

Eco-Support Graphics		1-800-724-7441

Audio Tapes
Green Island Spoken Audio Cooperative		1-800-438-0956

Automotive Products
Enviroline		1-800-466-1615

Bags
Ark Creations		1-800-646-2429
Best American Duffel (BAD)		1-800-424BAGS
ECO-BAGS		1-800-720BAGS

Bioremediation
Dixon Industries		1-800-4DIXON's

Business Opportunities
Greenway		1-800-966-1445
Kathleen Barber		1-800-872-7178

Children's Clothes & Gear
KidSystems		1800GOTOUGH
Tenax Corp	Baltimore MD	1-800-356-8495

Cotton Products
Janice Corporation		1-800-JANICES
GREENWEAR		1-800-934-1193

Dentistry - Environmental
Environmental Dental Association		1-800-388-8124

Deodorant - Personal
Deodorant Stone Mfg. Company		1-800-962-7863

Ecotourism

World:
American Wilderness Experience	1-800-444-0099
Creative Adventure Club	1-800-544-5088
Ecotours Worldwide	1-800-537-4025
Outer Edge Expeditions	1-800-322-5235
Overseas Adventure Travel	1-800-221-0814

Africa:
African Wildlife & Travel	1-800-432-9968
International Ventures Ltd.	1-800-727-5475
Solrep International Inc.	1-800-231-0985

Amazon:
International Journeys	1-800-622-6525
Chile, Antarctica, Patagonia:	
Latour	1-800-825-0825
Asia:	
Absolute Asia	1-800-736-8187
Central America:	
Costa Rica Experts	1-800-858-0999

Food - No Preservatives/Chemicals
Tamarind Tree   1-800-HFCTREE
ECO BAKERY   1-800-337-7ECO

Faucets - Plastic
NIBCO   1-800-642-5463

Fiber Fill - Batting & Stuffing
ECO-FIL   1-800-4-ECOFIL

Financial Investment - Environmentally & Socially Conscious
Calvert Social Investment Fund   1-800-368-2748
Capital Research & Management Co.   1-800-421-9900
Co-op America   1-800-424-2667
Dubuque Bank and Trust   1-800-397-3268
Good Money   1-800-535-3551
New Alternatives Fund, Inc.   Call Collect   1-516-466-0808
Parnassus Financial Management   1-800-999-3505
PAX - A Social Responsibility Fund   1-800-767-1729
Pioneer Fund   1-800-225-6292
Progressive Asset Management   Oakland CA   1-800-786-2998
Working Assets   1-800-223-7010

Fundraising
Green Earth Fundraising Programs   1-800-880-1915
Earth Fundamentals   1-800-50EARTH
The Nature Conservancy Adopt-An-Acre Program   1-800-628-6860
Project Earth   1-800-IM4ERTH

Futons - Non-allergenic
American Futons   1-800-821-8113

Futons - Recycled Fiber
Rising Star Futons   1-800-828-6711

General Merchandise - Earth Friendly
Environmental Supplies & Products   1-800-260-4320
Greenway   1-800-966-1445
Natural Lifestyle Supplies   1-800-752-2775
The Good Earth Catalog   1-800-392-6718
The Whole Earth Catalog   1-800-938-6657
The Woodstock Animal Rights Movement (WARM) Store   1-800889WARM

Gifts
HAWKWATCH   1-800-726-4295
Nonprofits   1-800-888-4741

Guidebooks - Travel
Lonely Planet   1-800-275-8555

Herbal Formulas
Rainforest Bio-Energetics   1-800-535-0503

Herbs/Supplements
Mountain Naturals of Vermont   1-800-992-8451

Homes & Houses - Energy Efficient
Deltec Homes   1-800-642-2508

Houseplants - Bioremediating
Half Environment Group Inc   1-800-285-5723
Nature's Air Filter   1-800-285-5723

Insurance - Health
National Association of Socially Responsible Organizations(NASRO) 1-800-638-8113

Insurance - Professional Liability
ECS Underwriting Inc. Exton PA   1-800-ECS-1414
Freberg Environmental Insurance   1-800-377-4152

Junk Mail - Stop
Georgetown Press   1-800-345-0096

Magazines
E The Environmental Magazine   1-800-825-0061
Garbage   1-800-825-6690

Media - Films & Video Tapes
Environmental Media   1-800ENVEDUC

Medical - First Aid Equipment
Atwater Carey Ltd.   Boulder CO   1-800-359-1646
Wilderness Medical Associates   Bryant Pond ME   1-800-742-2931

Musical Tapes
Centerstage Productions   1-800-553-4058

Lead
Lead Institute   1-800-532-3837
National Lead Information Center Clearinghouse   1-800-424-5323
National Lead Information Center Hotline   1-800-532-3394

Lights - Lighting
Environmental Lighting Concepts   1-800-842-8848

Meatless Meat
The Farm   1-800-695-2241

Mobiles
Skyflight Mobiles   1-800-766-8005

Nature Books
Northstyle   1-800-336-5666
King Tree   1-800-880-1636

Non-Profit Organizations & Foundations
Rainforest Preservation Foundation   1-800-460-RAIN

Nutritional Supplements
Energy for Life   1-800-927-2527

Office Supplies - Recycled
EcoTech Recycled Products	1-800-780-5353
Real Recycled	1-800-233-5335
National Copy Cartridge	1-800-822-5477

Outdoor Clothing
Carhartt	1-800-833-3118
REI	1-800-426-4840
Paper Towels - Household	
Valiant Paper	1-800-676-2754

Personal Care Products
Beauty Naturally	1-800-432-4323
Nadina's Cremes	1-800-722-4292
Oxyfresh USA Inc.	1-800-999-9551
The Body Shop	1-800-541-2535

Personals - Singles Network
Science Connection	1-800-667-5179

Pesticides - Exposure
Diana Fairechild, Flyana Rhyme, Inc.	1-800-LAG-TIPS

Publishers - Environmental Books
Cambridge University Press		1-800-872-7423
Executive Enterprises, Inc.	New York NY	1-800-831-8333
Harvard University Press		1-800-448-2242
Houghton Mifflin Company		1-800-304BOOK
Macmillan Publishing Co.		1-800-257-5755
MIT Press		1-800-356-0343
Noble Press		1-800-486-7737
Odonian Press		1-800-732-5786
Random House		1-800-733-3000
Regnery Gateway		1-800-462-6420
St. Martin's Press		1-800-288-2131
Simon & Schuster		1-800-233-2336
University of California Press		1-800-822-6657

Recycled Paper
Crestwood Recycled Paper Co.	1-800-525-3196
International Paper	1-800-242-2148
Message!Check Corporation	1-800-243-2565
Niagara	1-800-826-0431
Real Recycled (100%)	1-800-233-5335
Tree Free EcoPaper	1-800-775-0225

Recycled Printer Ribbons
NWORKS		1-800-547-4226
Printer Ribbon Recyclers	Lynnwood WA	1-800-336-4659

Recycling - Bags & Receptacles
The Bag Connection		1-800-622-2448
CSL & Associates	Ft Walton Beach FL	1-800-622-6069
Dyna-Pak Corp	Lawrenceburg TN	1-800-759-3962
Lincoln Recycling Services		1-800-989-7628
Windsor Barrel Works	Kempton PA	1-800-527-7848

Recycling - Market Information
American Recycling Market Directory   1-800-267-0707

Recycling - Promotion & Information
American Hospital Association   1-800-242-2626
John Lemmon Films (HenryCycle)   1-800-559-1953
Refuse Industry Productions Inc.   Grass Valley CA   1-800-576-3092
The Vinyl Institute's Environmental Resource Center   1-800-969-8469
Wildcat Composting   1-800-627-3954

Refrigerator - Energy Savers
The Conserve Group   1-800-814-3434

Shoes
VIBRAM   1-800VIBRAM7

Skin Care Products
Blackmores   1-800-433-9272
Rachel Perry   1-800-966-8888

Soap Products
Coastline Products   1-800-554-4112

Septic Systems
Septic Helper   Atlanta GA   1-800-533-2445

Telephone Companies
ATT Working Assets - Long Distance   1-800-CITIXEN

Toothpaste
Auromere Herbal Toothpaste   1-800-735-4691

Travel
Overseas Tours   1-800-323-8777
Paradise Management   1-800-367-5205

Tribal Crafts
Oxfam America   1-800-639-2141

Typewriter Ribbons - Reloaded
The Ribbon Factory   1-800-275-7422

Universities & Colleges - Environmental Programs
American Holistic College of Nutrition   1-800-659-2426
Chadwick University   1-800-767CHAD
Clayton School of Natural Healing   1-800-659-8274
Columbia Pacific University   1-800-227-0119
Columbia Pacific University   1-800-552-5522
P.C.D.I.   1-800-362-7070
University of Findlay Hazardous Materials Management Program
                         Findlay OH   1-800-521-1292
Vermillion Environmental Studies & The International Wolf Center   1-800-475-6666
Westbrook University   1-800-447-6494

Vegetable Nutrition
Pines Barley Grass   1-800-MYPINES

Video Duplication
HAVE Inc.   1-800-999HAVE

Video Tapes
Canyon Productions   1-800-644-4747
Madre Tierra (Mother Earth)   1-800-538-TAPE

Volunteering
Anasazi Crow Canyon Archeological Center   1-800-422-8975
USDA Soil Conservation Service Earth Team   1-800-THE-SOIL

Watches
Econiche International   1-800-425-1106

Washing Machines - Water Savers
ASKO   1-800-367-2444
White-Westinghouse   1-800-245-0600

Water Conservation - Information
Terra Firma Publications   1-800-345-0096

Water Purification
Carbon /Activated Carbon Filters:
Selecto, Inc.   1-800-635-4017
Valley Multi-Pure   1-800-767-4143

Weather Instruments
Davis Instruments   1-800-678-3669

Women's Undergarments
Travis Place Briefs   1-800-388-4101

Wool
Antizana   1-800-257-9407

Wholesale Merchandise
ESP. Inc.   1-800-886-5432

EMERGENCY RESPONSE & SPILL CONTROL - PRODUCTS AND SERVICES

American Boom and Barrier Corp	Cape Canaveral FL	1-800-843-2110
American Environmental Technology	Dingmans Ferry PA	1-800-828-7508
AIM USA	Houston TX	1-800-ASK4AIM
Allen-Bradley Response Center	Cedar Rapids IA	1-800-223-5354
American Health & Safety Inc.	Madison WI	1-800-522-7554
Brunswick Corp	Willard OH	1-800-243-7379
Carbtrol Corp	Westport CT	1-800-242-1150
Charles Houston Inc.	Jennings LA	1-800-325-8043
Containment Systems	Cocoa FL	1-800-282-4584
Dialogic Communications Corp	Franklin TN	1-800-969-3227
Douglas Engineering	Concord CA	1-800-533-8887
Duffs Environmental	Enterprise FL	1-800-523-4409
Du Pont Company	Plainwell IL	1-800-441-7515
Emergency Film Group	Plymouth MA	1-800-842-0999
Emergency Response Specialists	Birmingham AL	1-800-647-4ERS

Emergency Response Systems Inc.	Moreno Valley CA	1-800-599-5359
Emergency Technical Services Corp	Schaumberg IL	1-800-232-ETSC
EnviroPack Company	Arlington VA	1-800-423-8188
Environmental Container	Delafield WI	1-800-729-7137
Enviropact Inc.	Miami Lakes FL	1-800-234-0016
Four Seasons Environmental Inc.	Greensboro NC	1-800-868-2718
Illinois Chemical Corp	Wadsworth IL	1-800-624-7672
J. V. Manufacturing Company Inc.	Green Bay WI	1-800-334-9092
Life-Guard Inc.	Guntersville AL	1-800-323-2533
MSTX Ltd	Summit MS	1-800-992-8119
Matarah Industries Inc.	Milwaukee WI	1-800-222-4799
Micromedex	Denver CO	1-800-525-9083
National Sorbents Inc.	Hamilton OH	1-800-322-4144
Oil Mop Inc.	New Orleans LA	1-800-645-6671
Pacific Edge International	Seattle WA	1-800-822-1246
Perfex Corp	Poland NY	1-800-848-8483
Photographic Analysis Co.	Wayne NJ	1-800-524-0397
Reef Industries Inc.	Houston TX	1-800-231-2417
Sorbent Products Co. Inc.	Somerset NJ	1-800-333-7672
Spill Control Mfg & Supply Inc.	Bakersfield CA	1-800-472-9100
Spill Control Mfg & Supply Inc.	Bakersfield CA	1-800-356-0277
Texas Remediation Services Inc.	Brownwood TX	1-800-854-9275
Tim's Oil Recovery	St Petersburg FL	1-800-872-6715
Tri-State Transit Company	Joplin MO	1-800-872-8768
UltraTech International Inc.	Jacksonville FL	1-800-353-1611
Upright Inc.	St Louis MO	1-800-961-3711
Vacmasters of Denver Inc.	Denver CO	1-800-466-7825
Versatex Ecologic Absorbant Systems		1-800-409-4444
Waste Solutions	Sunnyvale CA	1-800-841-8628
World Safety Products Corp	Patchogue NY	1-800-441-4977

ENERGY CONSERVATION PRODUCTS -
WHOLESALE & MANUFACTURERS

ACI Energy Management Systems	Walnut Ridge AR	1-800-541-5590
Agway Energy Products	Greensburg PA	1-800-322-4929
Agway Energy Product of Chambersburg Pe	Chambersburg PA	1-800-548-5603
Air Enterprises	Newcastle ME	1-800-223-3774
Algie & Associates	Dana Point CA	1-800-442-7645
Alpha Energy Systems Inc.	Worcester MA	1-800-972-2743
American Sas of South Carolina	Pelzer SC	1-800-457-7274
Amerix	Fargo ND	1-800-232-4116
AMG Energy Inc.	Hackensack NJ	1-800-752-5050
Barber Watterson Co.	Louisville KY	1-800-722-5040
Bier TM & Associates	Glen Cove NY	1-800-862-3674
Bonar Engineering	Jacksonville FL	1-800-621-8168
Cain Industries	Germantown WI	1-800-558-8690
California Energy Designs Inc.	La Canada CA	1-800-426-5982
Carl Heinrich Company	Cambridge MA	1-800-225-1945
Chicago Stell Tape	Watseka IL	1-800-435-1859
Command Systems Inc.	Columbus GA	1-800-542-4840
Condyne Technology & Information	Longwood FL	1-800-643-6374
Continental Diversified	Redwood City CA	1-800-346-1150
Control Solutions Inc.	Lebanon OH	1-800-356-3627

Cool Curtain	Cucamonga CA	1-800-854-5719
Cool Curtain	Cucamonga CA	1-800-662-8872
Curtron Industries Inc.	Catskill NY	1-800-833-5005
Davis Controls	Rolling Meadows IL	1-800-322-5712
Demand Side Resources	Kalamazoo MI	1-800-362-2525
Demand Side Resources	Plainfield IN	1-800-821-2307
Dencor	Denver CO	1-800-392-2690
DMC Services	Hammonton NJ	1-800-447-2177
Earlwood Technologies	Norristown PA	1-800-233-5193
Earth Guard Environmental Inc.	Henderson NV	1-800-559-0291
Ecology Enterprises	Oelwein IA	1-800-848-8931
Econs Inc.	Redmond WA	1-800-828-8440
EDC Technologies	Santa Fe Springs CA	1-800-634-3328
Enercom	Turners Falls MA	1-800-632-5947
Energy Compliance Systems	San Jose CA	1-800-824-2252
Energy Conservation Technologies	Cottage Grove OR	1-800-432-5560
Energy Management Systems	Towson MD	1-800-541-5740
Energy Management Systems	Towson MD	1-800-537-3911
Energy Systems Inc.	Mobile AL	1-800-872-3230
Energy Technology Lab	Modesto CA	1-800-344-3242
Engineered Energy Services	Jackson TN	1-800-428-1175
Engineering Resources Inc.	Chicago IL	1-800-258-6771
Environmental Control	Marietta GA	1-800-542-4507
Fennicks Independent Energy	Ravenna OH	1-800-336-6425
1st National Warranty Corp.	Greenwood IN	1-800-526-4665
Flourescent Maintenance Service	Jackson MS	1-800-345-6986
G & H Service	Norristown PA	1-800-362-0353
Garrett Technology	Norcross GA	1-800-833-0723
Gen Pro	Sioux City IA	1-800-722-3791
Genotec	Columbus OH	1-800-243-3599
Gill Construction, Energy Management Di	Woodland Hills CA	1-800-446-4848
Glo-Tech Ltd.	New York NY	1-800-424-2883
Greater Memphis Distributors	Memphis TN	1-800-245-8155
Hendee Enterprises	Houston TX	1-800-231-7275
Hydro Engineering	Salt Lake City UT	1-800-247-8424
ICSS-Illuminating Consultants Service &	Montgomery Center VT	1-800-356-0277
International Energy Savers Inc.	Tempe AZ	1-800-962-7283
Janeco Inc.	Tempe AZ	1-800-962-3827
JM Basa Enterprises Energy Release	Carson City NV	1-800-325-5673
Kelar Controls Inc.	San Jose CA	1-800-538-3098
Kingman Industries	Santa Ana CA	1-800-854-7214
Koala Energy Systems Inc.	St. Louis MO	1-800-831-6188
Land Tec (Alternative Energy)	City Of Commerce CA	1-800-821-0496
M & J Energy Control Systems Inc.	Warner Robins GA	1-800-541-7933
Martin Processing	Arlington TX	1-800-762-3328
Midwest Energy Control Products	Long Grove IL	1-800-742-8080
Neotronics Of North America	Gainesville GA	1-800-535-0606
Niagara Conservation Corp.	Flanders NJ	1-800-831-8383
North Carolina Energy Division	Raleigh NC	1-800-662-7131
Pacific Energy & Gas Company	Newport Beach CA	1-800-233-2464
Pacific Gas & Electric Co.	San Ramon CA	1-800-468-4743
Monthly Journal of Ferc Daily Activitie	Mt. Rainer MD	1-800-538-1671
Poly Mfg. Inc.	Dallas TX	1-800-628-6372
Practical Energy Group	Huntington NY	1-800-562-8929
Prime Serivce & Supply	Kirkland WA	1-800-772-9912
Pro-Tech Intl.	Jupiter FL	1-800-345-4434

Proseal Products	St. Paul MN	1-800-233-1117
R. L. Adams Inc.	Hope Mills NC	1-800-424-2808
Remtec Systems	Gardena CA	1-800-421-2662
Resources Conservation Inc.	Greenwich CT	1-800-243-2862
Rise	Providence RI	1-800-422-5365
Sea Kay Energy	E Lyme CT	1-800-344-5899
Shay Corporation	Hemet CA	1-800-221-6684
Solar Master Film Corporation	Pennsauken NJ	1-800-257-0448
Southland Shade	Costa Mesa CA	1-800-638-7368
Steamsphere	Jerseyville IL	1-800-551-3361
Stimpson Intl. Inc.	Akron OH	1-800-382-7595
Stitt Energy Systems	Rogers AR	1-800-367-7374
Tech S	Livonia MI	1-800-228-3247
Telemark Inc.	Little Rock AR	1-800-874-5753
Thermal Art Inc.	Paramount CA	1-800-628-4328
Thermco Intl.	Scottsdale AZ	1-800-858-8911
Thermwell Products	Paterson NJ	1-800-526-5265
UEC	Wylie TX	1-800-833-8937
Unenco-Unisec	San Leandro CA	1-800-227-0452
Utility Reduction Consultants	Crown Point IN	1-800-423-6552
Utilysis Inc.	Tucker GA	1-800-443-5732
Viewtech	Columbus OH	1-800-545-0263
Vision Impact	Youngstown OH	1-800-321-4689
World Mfg. Inc.	Costa Mesa CA	1-800-722-3539

ENERGY MANAGEMENT & CONSERVATION CONSULTANTS

A-Tec Energy Corp	Des Moines IA	1-800-798-1704
Accu Temp	Indianapolis IN	1-800-854-4835
Architectural Energy	Boulder CO	1-800-921-4653
DMC Services	Brick NJ	1-800-832-0035
Earth Guard Environmental Inc.	Henderson NV	1-800-559-0291
Energy Conservation Concepts	Deerfield Beach FL	1-800-275-3318
Fred Davis Corporation	Medfield MA	1-800-497-2970
Genesis Energy Development	Napa CA	1-800-999-3696
Gill Construction Energy Management Div	Woodland Hills CA	1-800-446-4848
Gold Energy Designs		1-800-535-0191
Lippincott Energy Services	Lima OH	1-800-257-7918
McLaughlin Intl	Palm Harbor FL	1-800-231-4664
National Energy Management Institute	Lakewood CO	1-800-333-6364
Pacific Northwest Laboratory Building E	Richland WA	1-800-270-2633
Planergy	Houston TX	1-800-527-7648
Power Conditioning Systems	Long Valley NJ	1-800-452-0469
SSP Consultants	Aurora CO	1-800-279-1057
Synertech Systems	Syracuse NY	1-800-955-3656
Utility Rebate Corp.	Austin TX	1-800-624-3377
Westcoast Gas Services USA Inc	Traverse City MI	1-800-947-9411

ENGINEERS

21st Century Engineering	Ft Lauderdale FL	1-800-338-0398
A Shannon Engineering	Ft Worth TX	1-800-432-1777
Academy Consulting	Palm Desert CA	1-800-696-4436

Ace Engineers	TAMARAC FL	1-800-548-0002
Ace Engineers	Haddonfield NJ	1-800-548-0002
Acoustical Consultants	Seattle WA	1-800-843-4524
ACR Engineering	Austin TX	1-800-443-6878
Adcot Consulting	Merritt Island FL	1-800-367-4125
Aecl Technologies	Rockville MD	1-800-872-2325
AECL Technologies Inc.	Atlanta GA	1-800-842-2325
Aegis Engineering	S Walpole MA	1-800-247-8910
Aid Consulting Engineers Inc	Dallas TX	1-800-433-8984
Air Control Engineering	Thousand Oaks CA	1-800-548-0311
Alex Dixon Inc.	Randallstown MD	1-800-922-0043
Alico Engineers & Appraisers Inc	Birmingham MI	1-800-545-0788
Allen & Wright Services Inc.	Atlanta GA	1-800-354-6838
Allied Engineers & Surveyors Inc.	Orlando FL	1-800-433-4875
Alpha Engineering Service	Beckley WV	1-800-252-4131
Amada Engineering & Serive	La Mirada CA	1-800-338-2374
Amber Engineering	Goleta CA	1-800-232-6237
American Chemistry Society STN Help Desk	Chemical Abstracts	1-800-848-6533
American Nuclear Society	La Grange IL	1-800-682-6397
American Piling And Foundation Corporat	Victor NY	1-800-532-2732
American Standards Testing Bureau	New York NY	1-800-221-5170
Analysis & Technology Inc.	New London CT	1-800-824-3289
Anco Engineers Inc.	Culver City CA	1-800-932-5515
Anzelmo & Associates-Engineers/Land Sur	New Haven KY	1-800-634-3775
Apex Cadd Services	Youngstown OH	1-800-821-2739
Applied Energy Systems	Tulsa OK	1-800-752-0604
Ardaman & Associates Inc.	Orlando FL	1-800-432-3143
Arora & Associates PC-Consulting Engine	Lawrenceville NJ	1-800-628-2596
Artec	St Louis MO	1-800-622-1454
Arthur G Lewis Mgmt. Engineers	Grayton CA	1-800-255-9443
ASC Professionals	Roseville MI	1-800-327-4260
ASM International	Materials Park OH	1-800-336-5152
Babet Engineering	Pasedena TX	1-800-932-2156
Bae Structural Engineering	Las Vegas NV	1-800-524-6642
Baker TSA	Coraopolis PA	1-800-553-1153
Balchuck Automotive	Pueblo CO	1-800-627-0455
Bartech	Madison Heights MI	1-800-824-2962
BASF Engineering Plastics Customer Serv	Bridgeport NJ	1-800-527-8324
BCM Engineers	Pittsburgh PA	1-800-447-3379
BCM Engineers Inc.	Mobile AL	1-800-451-9008
Becht Engineering	Liberty Corner NJ	1-800-772-7990
Beehive Engineering Inc	Orinda CA	1-800-367-0764
Beehive Engineering Inc.	Moraga CA	1-800-367-0764
Belcan Corporation	Cincinnati OH	1-800-543-1154
Belcan Corporation	Cincinnati OH	1-800-543-4543
Belco Engineering	Pasadena CA	1-800-833-1833
Benchmark Engineering & Surveying Inc.	Toledo OH	1-800-245-4133
Bendix Field Engineering Corp.	Cocoa Beach FL	1-800-762-9191
Biocybernetics Lab Inc.	Danvers MA	1-800-782-3343
Black & Veatch	Greenville SC	1-800-922-3451
Blue Ridge Engineering	Atlanta GA	1-800-258-0992
Brice Petrides-Donohue	Waterloo IA	1-800-772-2028
Brunkhorst Engineering Consultants Inc.	Port Clinton OH	1-800-442-4224
Bryant David & Associates	Melbourne FL	1-800-422-4768
Bryant David & Associates	Melbourne FL	1-800-422-4768
BSE Consultants	Melbourne FL	1-800-523-4273
Buck Seifert & Jost	Paramus NJ	1-800-245-1105
Building Systems RX	Redmond OR	1-800-821-3778

Burdett Daniel S PE PC	New York NY	1-800-982-9876
Butler Fairman & Sevfert Engineers	Carmel IN	1-800-553-0863
C & O Associates Inc. Technical Placeme	Richmond VA	1-800-872-9464
C & W Project Consultants	Orange Park FL	1-800-262-7718
C H & A Corp	San Diego CA	1-800-998-9588
C H & A Engineer Consultants	New Orleans LA	1-800-437-7608
C P Test Services	Harrison NJ	1-800-282-5269
C-Water Engineers	Kinston NC	1-800-523-7973
C-Water Engineers	Kinston NC	1-800-523-7973
Cabe Associates Inc	Dover DE	1-800-542-7979
Cable Testing Service Inc	North East MD	1-800-624-1600
Camp Dresser & McKee Inc.	Cambridge MA	1-800-243-2677
Cannon Construction	Mcdonald TN	1-800-441-7135
Capital Communications Industries	Olympia WA	1-800-426-8664
Capital Communications Industries	Olympia WA	1-800-562-6006
Capsule Environmental Engineering Inc.	St Paul MN	1-800-328-8246
Carlander & Associates	Casselbury FL	1-800-457-6444
Cataract Inc.	Newtown PA	1-800-523-8960
CCS Control Systems	Victorville CA	1-800-669-2271
Cedar Corp.	Menonomie WI	1-800-472-7372
Central Florida Testing Labs	Largo FL	1-800-248-2385
Certified Structure & Foundation	Melbourne FL	1-800-543-8968
CH2 M Hill	Herndon VA	1-800-242-6445
Clark Charles N Assoc Ltd	Laurel MS	1-800-898-0162
Coherent Technologies	Missouri City TX	1-800-874-6820
Cole & Associates	Shelbyville TN	1-800-782-6302
Columbia Engineering	Wenatchee WA	1-800-721-8052
Construction Technology Lab	Skokie IL	1-800-522-2285
Construction Technology Laboratories In	Skokie IL	1-800-522-2285
Coon Engineering Inc.	Oklahoma City OK	1-800-338-1326
Corrosion Engineering & Research	Concord CA	1-800-447-2032
Crum & Associates	Omaha NE	1-800-243-3613
CSC Professional Services Group Inc	Rockville MD	1-800-272-0033
CTR Technical Services Inc	Arlington TX	1-800-524-9328
D K Enterprises-Electronics Engineering	N Hollywood CA	1-800-451-2012
Danco Engineering	El Cajon CA	1-800-522-5141
Daniel Plant Engineers	Greenville SC	1-800-522-5647
DC Consulting & Equipment	Afton OK	1-800-854-1485
Dewild Grant Reckert & Associate	Sioux Falls SD	1-800-446-4157
Diamond Group	Odessa DE	1-800-220-4060
Dl Engineering	Odessa TX	1-800-725-7250
Dodd Pacific Engineering Inc.	Seattle WA	1-800-621-7300
Dreher Charles R	Miami FL	1-800-428-3556
DuPont		1-800-231-0998
Dynamic Analysis	St Leawood KS	1-800-842-0616
E M A	Oxford MS	1-800-364-3898
Earth Engineering Sciences	Baltimore MD	1-800-346-5271
Earth Lab Inc.	Glen Burnie MD	1-800-343-4041
Ebasco Services BBS	Cary NC	1-800-582-3383
Edwards and Kelcey Inc.	Livingston NJ	1-800-253-9527
Eldorado Engineering Co	New Orleans LA	1-800-500-5518
Elektromekanik Intl	S Plainfield NJ	1-800-353-5876
Elrod & Co.	Murfreesboro TN	1-800-553-5111
EMD Technologies Inc.	Bensenville IL	1-800-551-3755
Engineered Pipeline Systems Inc	Odessa TX	1-800-221-0836
Engineering Associates	Syosset NY	1-800-458-4681
Engineering Co Inc	Boston MA	1-800-343-0863
Engineering Design & Testing	Columbia SC	1-800-338-3227

Engineering Design & Testing Corp	Houston TX	1-800-452-1741
Engineering Design & Testing Corp	Jacksonville FL	1-800-851-7869
Engineering Design & Testing Corp	Jacksonville FL	1-800-851-7869
Engineering Technology	Austin TX	1-800-367-4238
Enprotec	Arlington TX	1-800-391-0701
Enreco Engineering Group	Lubbock TX	1-800-972-8467
Environmental Geotechnical Specialists	Tallahassee FL	1-800-332-1253
Environmental Resource Associates Inc	Warwick RI	1-800-338-7398
Ercikson Engineering	Minneapolis MN	1-800-545-8020
Esmond Engineering Inc	College Station TX	1-800-444-7640
Estes Engineering	San Francisco CA	1-800-822-7173
Etc Engineers Inc.	Little Rock AR	1-800-472-1715
F E S Ltd Publishing	Stuart FL	1-800-337-5831
Fatigue Concepts	El Dorado Hills CA	1-800-342-7225
Fleet Charles H & Associates	Richmond VA	1-800-439-7504
Fluid Systems Engineering Inc.	Morristown NJ	1-800-854-5536
Forsgren Perkins Engineering	Rexburg ID	1-800-331-7548
Forsgren Perkins Engineering	Salt Lake City UT	1-800-826-9304
Franklin Engineering	Panama City FL	1-800-433-8539
Franklin Engineering Co. Inc.	N Little Rock AR	1-800-272-5665
Franklin Engineering Co. Inc.	Memphis TN	1-800-238-7500
FTI Corp	Annapolis MD	1-800-334-5701
G & G Consultants of Mississippi Inc.	Starkville MS	1-800-322-7731
G P U-Power Team	Parsippany NJ	1-800-882-9229
Gasser Associates	Olney MD	1-800-654-4119
Gaumer's Chasis Engineering	Chambersburg PA	1-800-521-0425
Geotech Services Inc.	Brookfield CT	1-800-633-4436
German & Milne Inc.	Houston TX	1-800-822-8119
Giles Engineering Associates	Anaheim CA	1-800-445-3723
Gilmore & Associates	New Britain PA	1-800-344-5667
GLA Engineering Inc.	Natick MA	1-800-281-1187
Glace & Radcliff	Maitland FL	1-800-600-4778
Gladwin Engineering Systems	Columbiana OH	1-800-452-3946
Gleason T A Associates	Cincinnati OH	1-800-842-7114
Glenco Engineering	Buffalo Grove	1-800-562-2543
Golden Circuits	Golden CO	1-800-638-5287
Gooch John Associates	McLean VA	1-800-524-6624
Haag Engineering Co.	Carrollton TX	1-800-442-0814
Haag Engineering Co.	Carrollton TX	1-800-527-0168
Haag Engineering Co.	Houston TX	1-800-635-0116
Haag Engineering Co.	New Orleans LA	1-800-631-5192
Hallibutron NUS Environmental Group	Downers Grove IL	1-800-742-5409
Hamlin & Associates Inc - Worldwide	EngDuncan OK	1-800-975-0600
Hansen & Lupien Corp	N Troy VT	1-800-639-4381
Hanson Thomas C Co Inc	Savage MN	1-800-637-5014
Hanson Thomas C Co. Inc.	Savage MN	1-800-637-5014
Hendershot Jim	Grants Pass OR	1-800-343-9600
Hensley Schmidt Inc.	Chatanooga TN	1-800-826-4960
Hoechst Celanese	Charlotte NC	1-800-242-6222
Hope Engineers	Benton AR	1-800-700-7676
Hovde Engineering	Ridgecrest CA	1-800-982-4270
Hudson Intl	Columbia MD	1-800-787-7844
Human Factors Engineering Inc.	Lake Bluff IL	1-800-433-9033
Hunsaker & Associates/San Bernadino Inc	San Bernadino CA	1-800-637-6167
Iesco	San Pedro CA	1-800-944-3726
Indiustry Consulting Engineers	Portland OR	1-800-346-6195
Industrial Communication Engineering Lt	Indianapolis IN	1-800-423-2666
Industrial Design Ch2M Hill	Tallahassee FL	1-800-235-8997

Industry Consulting Engineers	Portland OR	1-800-346-6195
Inmicro	Miami FL	1-800-643-1738
Innovative Logistic Techniques	McLean VA	1-800-325-0658
Institute of Advanced Manufacturing Sci	Cincinnati OH	1-800-345-4482
Institute Of Advanced Mfg Science	Cincinnati OH	1-800-345-4482
Intech Engineering	Marlton NJ	1-800-521-6225
Intera Info Technology Inc.	Denver CO	1-800-972-2953
Interconnect Design Inc.	Norcross GA	1-800-437-7927
Intl Quality	Aurora CO	1-800-500-0651
J E O'Toole Engineering Co. Inc.	Birmingham AL	1-800-368-2671
J S Smith Consulting	Joplin MO	1-800-368-8429
Jaca	Ft Washington PA	1-800-292-2510
Jenkins & Charland	Ft Myers FL	1-800-394-5454
Jenlynn Intl Inc	Boulder CO	1-800-822-8912
Jennings, Thomas A.	Atascadero CA	1-800-252-6793
Jennlyn International Inc.	Boulder CO	1-800-822-8912
JHR Associates	Las Vegas NV	1-800-252-6454
Johnson Service Group	Atlanta GA	1-800-841-0927
Johnson Service Group	Des Plaines IL	1-800-323-4987
Jones John David & Associates	Cuyahoga Falls OH	1-800-336-5352
Keefer Veroski Associates	Orlando FL	1-800-355-1582
Kelley Gidley Blair & Wolfe	Charleston WV	1-800-304-5429
Kharbanda	Miami FL	1-800-236-8729
Kipcon Inc.	N Brunswick NJ	1-800-828-4118
KLH Engineering Group Inc.	San Ramon CA	1-800-452-7273
Kovacs Structural Engineering Inc.	Port Charlotte FL	1-800-526-9111
Lagniappe Enterprises Inc.	Ballwin MO	1-800-243-6195
Lanco Engineer Co.	Birmingham AL	1-800-872-9161
Landmark Engineering Inc.	New Castle DE	1-800-368-7483
Laughter Austin & Associates	Hendersonville NC	1-800-858-5263
Law Engineering	Charlotte NC	1-800-672-6001
Leighton & Associates Inc.	Irvine CA	1-800-253-4567
Liesch B A Associates - Environmental	EPlymouth MN	1-800-338-7914
Liesch B A Associates - Environmental	EAlbert Lea MN	1-800-338-7917
Lockwood Greene Engineers Inc.	Spartanburg SC	1-800-845-3302
Loss Research & Analysis	Greenville TX	1-800-442-1774
Lowery John L & Associates	Baton Rouge LA	1-800-535-8375
Lumos & Associates	Fallon NV	1-800-531-7155
M & L Engineering	Henderson KY	1-800-247-8576
M & T Engineering	Veradale WA	1-800-228-7599
M A day & Associates Inc.	Poughkeepsie NY	1-800-841-0878
Mabbett & Associates	Bedford MA	1-800-877-6050
Madsen Jacobs Kneppers Inc.	Englewood CO	1-800-822-6624
Mansur Daubert Strella	Tulsa OK	1-800-253-3647
Manteq	Lake Jackson TX	1-800-633-4473
Marathon Technical Services	Summersville WV	1-800-524-8210
Mark 3 Construction	Troy AL	1-800-933-6275
Matco Associates	Pittsburgh PA	1-800-221-9090
Materials Engineering & Technology Inc.	E Rutherford NJ	1-800-331-3009
May Engineering Lab Inc	Sioux Falls SD	1-800-321-7654
McClure Engineering Associates Inc.	E Moline IL	1-800-346-8848
McDonnell Douglas Technical Services Co	Long Beach CA	1-800-352-3737
McMillen Engineering	Hopwood PA	1-800-242-1244
Meagan Corp The (Consulting)	Allentown PA	1-800-344-6275
Meehanite Worldwide Corp	Beaufort SC	1-800-423-0992
Merrick & Co.	Aurora CO	1-800-544-1714
Microdimensions Inc.	Mentor OH	1-800-423-7252
Mid-Columbia Engineering Inc	Richland WA	1-800-366-7028

Mid-State Associates	Baraboo WI	1-800-362-4505
Midcolumbia Engineering	Richland WA	1-800-322-6706
Midwest Technical Inc.	Kingsport TN	1-800-242-1517
Miller Process Coating Co	E Pittsburgh PA	1-800-742-9170
Mobilab/Alice Kupper PE	S Amboy NJ	1-800-545-7645
Moffet Larson & Johnson MLJ	Falls Church VA	1-800-523-3117
Monroe & Newell Engineering	Avon CO	1-800-466-1660
MSA & Associates Simin Razavian & Mehrd	La Jolla CA	1-800-459-5665
MYR Engineering	Hampden MA	1-800-257-3383
National Environmental Contractors	Cumming GA	1-800-847-0104
Nelson Surveying & Engineering-Consulta	Ashland WI	1-800-682-9780
NES Inc.	Braintree MA	1-800-848-7795
Network Design Engineer Of Little Rock	Little Rock AR	1-800-221-0911
New England Electrical Testing Engineer	Walpole MA	1-800-345-9205
New Horizons Technologies	Jackson MI	1-800-858-8982
New Horizons Technologies Inc.	Jackson MI	1-800-242-1632
Nielson Geo Technical	Citric Heights CA	1-800-255-9583
Noramco Engineering	Hibbing MN	1-800-262-1093
North East Engineering Consultants	Springfield MA	1-800-557-8527
North Engineering	Ann Arbor MI	1-800-524-8461
Northern Engineering Corp	Detroit MI	1-800-841-8878
Northwestern Territories Inc.	Port Angeles WA	1-800-654-5545
Notestein Well Drilling & Engineering C	Appomattox VA	1-800-462-9355
NOVA Engineering Inc.	Tulsa OK	1-800-727-6005
NUS Corp.	Gaithersburg VA	1-800-368-2755
NUS Corp.	Gaithersburg VA	1-800-233-7299
Nyhus Engineering	Henning MN	1-800-446-5038
Oasis	Brookfiled WI	1-800-736-0350
Oasis	Rochester NH	1-800-800-5335
Online-Consulting Engineers	Fruitland NM	1-800-841-6805
Owen & White	Baton Rouge LA	1-800-535-8342
P & C Research	Tulsa OK	1-800-545-6314
P O D Associates	Albuquerque NM	1-800-664-5559
Pacific Design Engineer	Beaverton OR	1-800-445-9039
Parallel Assoc	Stratham NH	1-800-200-8888
PCI Energy Services Inc.	Shrewsbury MA	1-800-841-7707
Pearce Feller Kanodia & Associates	Akron OH	1-800-654-5874
Pearlson Professional Buildings	Miami FL	1-800-359-0571
Plant Layout Services	Denver CO	1-800-624-2969
Pollak & Skan Inc.	Kalamazoo MI	1-800-392-3081
Post Buckley Schuh Jernigan	Miami FL	1-800-445-7275
Praxair/Linde	Danbury CT	1-800-PRAXAIR
Precision Automation	Cherry Hill NJ	1-800-637-6553
Prism Systems	Mobile AL	1-800-582-7747
Progressive Architecture Engineering &	Brighton MI	1-800-544-1983
Project Planning Group	Ocean NJ	1-800-551-0111
Protocol	Mt. Olive NJ	1-800-346-7910
Proudfoot Associates	Toledo OH	1-800-524-2877
PSI Environmental Materials	Birmingham AL	1-800-533-9561
Quality Inspection Services Co.	Columbia MD	1-800-572-0533
Quest Inc.	Metaire LA	1-800-453-5611
Questech Inc.	Falls Church VA	1-800-336-0354
R & R International	Akron OH	1-800-537-8084
R W Armstrong & Associates	Knoxville TN	1-800-453-6960
Raisbeck Engineering Hdqrs Inc	Seattle WA	1-800-537-7277
Ralph M. Parsons Employment	Pasadena CA	1-800-237-2584
Randall Engineering	Milton-Freewater OR	1-800-554-3928
RCG Research-Engineer Consultants	Noblesville IN	1-800-442-8272

Reece Anderson & Associates	Houston TX	1-800-245-4161
RH2 Engineering	Olympia WA	1-800-722-8052
Rockford Engineering Services	Sunol CA	1-800-848-3781
Roy M Benjamin PE	Brigantine NJ	1-800-882-8797
Scheiner Corp	Hixon TN	1-800-531-9264
Scientific Software	Houston TX	1-800-238-0116
Seguro Project Services Inc	Denver CO	1-800-457-5790
Self Propelled	Marlboro CT	1-800-223-7353
Semcor	Chesapeake VA	1-800-842-5408
Senior Engineering Co.	King of Prussia PA	1-800-523-0480
Sensornetics	Aloha OR	1-800-248-2456
Servo Tech Engineering	Belleville MI	1-800-737-8538
Shacoh USA	Walled Lake MI	1-800-252-6793
Shanta Engineering	Cerritos CA	1-800-831-9032
Sharp George Inc. Marine	Arlington VA	1-800-424-7003
Shefchick Associates	Sunnyvale CA	1-800-321-4866
Sheppard Crane & Associates	Stafford TX	1-800-422-7252
Simone's Enterprises Inc	Miamisburg OH	1-800-435-8435
Simpson Gumpertz & Heger	San Francisco CA	1-800-426-4500
Smith Seckman Reid	Nashville TN	1-800-545-6732
Southwestern Engineer	King of Prussia PA	1-800-533-8971
Specialty Measurements	Pittstown NJ	1-800-772-3570
Star Plus Softwear Inc	Oakland CA	1-800-446-7848
Statco Engineers & Fabricators	Huntington Beach CA	1-800-421-0362
Steinman Boynton Gronquist & Birdsall	New York NY	1-800-253-2515
Stevens Engineers Inc	Hudson WI	1-800-822-7670
Stone and Webster	Houston TX	1-800-772-2668
Sunbelt Optics Inc	Cantonment FL	1-800-331-2358
Swift Engineering	Colfax CA	1-800-382-0834
Synergy Engineering Technologies	Glendale AZ	1-800-797-3647
Syska and Hennessy	New York NY	1-800-328-1600
Sysology Corp	Whiting IN	1-800-642-4312
System Engineering & Laboratory	Tyler TX	1-800-432-4669
System Engineering & Laboratory	Tyler TX	1-800-624-0905
Systems Control Technology	Palo Alto CA	1-800-227-1910
TAD Telecommunications Services	Atlanta GA	1-800-843-6910
Talbert Cox Associates	Charlotte NC	1-800-822-1502
Taylor Richard Engineering Inc.	Los Angeles CA	1-800-252-0070
Technical Engineering & Designs	Lolo MT	1-800-358-9550
Technical Services	Vancouver WA	1-800-821-4587
Tesco Inc	Mobile AL	1-800-467-6103
Texas Engineering Extension Service	College Station TX	1-800-252-2420
TFL Inc.	Houston TX	1-800-828-5002
Timco Engineering	Miami FL	1-800-233-3587
TMI Engineering Inc.	New York NY	1-800-225-5864
Tracor Applied Science Inc.	Rockville MD	1-800-638-8512
Travis Pruitt and Associates	Tucker GA	1-800-362-6593
TT & I	Winnetka CA	1-800-884-3647
United Engineers	Norwalk CT	1-800-243-5455
Universal Ensco	Roswell GA	1-800-742-5174
US Army Corps Of Engineers	Champaign IL	1-800-252-7122
V & A Consulting Engineers	Oakland CA	1-800-824-8404
Vibra Tech Engineers	Hazelton PA	1-800-233-6181
Volt Technical Services	Blue Bell PA	1-800-523-6609
Von Otto & Bilecky	Stewartstown PA	1-800-874-5032
Votechnics	Portland ME	1-800-882-1616
W W Engineering & Science Inc.	Grand Rapids MI	1-800-688-9828
Walsh Estimating Service	E Stroudsburg PA	1-800-448-1167

Watkins Partnership The	Bowie MD	1-800-310-9785
Wehling Engineers	Beatrice NE	1-800-622-2905
White Barry & Associates	Columbus TX	1-800-526-3787
Willbanks Corp	Denver CO	1-800-545-5403
Wilson & Co.	Salina KS	1-800-432-7921
Witman & Howard	Wellesley MA	1-800-342-8811
Woodard & Curran	Bangor ME	1-800-624-6767
Woodard & Curran Inc.	Portland ME	1-800-426-4262
Woodward-Clyde Group Inc.	Denver CO	1-800-776-3296
X P Software	Tampa FL	1-800-883-3487
Yoh Hi	Springfield MA	1-800-342-5964
York Power Systems-Diesel Engines	Houston TX	1-800-752-4996
Youchak & Youchak Inc.	W Palm Beach FL	1-800-622-9515

ENVIRONMENTAL & ECOLOGICAL SERVICES

1st Strike Environmental	Roseburg OR	1-800-447-3558
A & N Environmental Answers	Pacific Palisades CA	1-800-219-8243
A 2000	Ho Ho Kus NJ	1-800-446-6050
Abasco	Houston TX	1-800-242-7745
Absolute Standards	Hamden CT	1-800-368-1131
Accuchem Inc	Everson WA	1-800-636-7645
Advanced Environmental Solutions	Folcroft PA	1-800-688-0883
African Wildlife Foundation	Washington DC	1-800-344-8875
Air Sciences & Engine Technology	Houston TX	1-800-982-0584
Allstates Environmental Services Inc.	St Charles MO	1-800-851-5446
American Environmental Technologies Inc	Bethel CT	1-800-562-7611
American Oceans Campaign	Santa Monica CA	1-800-862-3260
Anti-Pest Co	Manhattan KS	1-800-432-8234
Appalachian Lab Inc	Forked River NJ	1-800-564-3100
Aptus Inc	Lakeville MN	1-800-328-4061
Associated Environmental Services	Bear DE	1-800-550-2371
B & E Environmental Systems	Amarillo TX	1-800-375-8401
Bacteria Concepts	Downers Grove IL	1-800-343-5255
Basic Concepts	Fremont NH	1-800-249-5261
Ben Meadows Company, The	Atlanta GA	1-800-241-6401
Bio-Medical Waste Services	Miami FL	1-800-281-3359
Bioclean Environmental Inc	Monument Beach MA	1-800-339-9728
Bionomics	Kingston TN	1-800-578-6513
C & W Resources Intl	Bellaire TX	1-800-538-5897
C C Lynch & Associates	Pass Christian MS	1-800-333-2252
California Business Environmental Assis	Anaheim CA	1-800-662-2322
Caray-All Shopping Bags Greensboro Nc	Oolitic IN	1-800-253-7225
Carbon Group The	Houston TX	1-800-230-9461
Carlo Environmental	Clinton Township MI	1-800-238-9042
Cascade Environmental	Federal Way WA	1-800-793-5359
Cedar Environmental Services	Fishkill NY	1-800-492-3327
Chemical Solutions International	Pearland TX	1-800-424-4804
Chemical Wate Management Inside Sales	Oak Brook IL	1-800-541-5511
Chesapeake Regional Information Service	Richmond VA	1-800-662-2747
Clarke Mosquito Control Products Inc.	Roselle IL	1-800-323-5727
Crosby Lab	Anaheim CA	1-800-327-6729
Data Collection Services	Cave Junction OR	1-800-592-3282
Dewey Pest Control	Bakersfield CA	1-800-653-6606
Dewey Pest Control	Colton CA	1-800-653-6626
Dewey Pest Control	Baldwin Park CA	1-800-653-6636
Dewey Pest Control	Colton CA	1-800-697-9780

Dewey Pest Control	National City CA	1-800-697-9784
Dewey Pest Control	Laguna Hills CA	1-800-697-9785
Dewey Pest Control	Oroville CA	1-800-769-6003
Dewey Pest Control	Thousand Palms CA	1-800-769-6004
Dewey Pest Control	San Jose CA	1-800-769-6008
Dewey Pest Control	San Marcos CA	1-800-998-0677
Dewey Pest Control	Pasadena CA	1-800-998-0681
Diversified Environmental Recovery Serv	Jackson MO	1-800-242-9117
DLS Lab Inc	Oklahoma City OK	1-800-831-8809
Dynamac	Millington TN	1-800-346-6732
Dynamic Environmental D E C Sales	Holtsville NY	1-800-770-5777
E & K Hauling	Sheboygan WI	1-800-688-4005
Eagle Instruments Inc.	Stone Mountain GA	1-800-647-8289
Earth 98	Van Nuys CA	1-800-327-8498
Earth Action Recyclers	Gainesville FL	1-800-327-8422
Earth Remediation Services	Branford CT	1-800-457-6713
Eason & Smith Enterprises	Oklahoma City OK	1-800-441-6672
Eco Expo	Canoga Park CA	1-800-471-4711
Ecology Alternative Inc	Freehold NJ	1-800-551-7645
Ecosource	Atlanta GA	1-800-864-2737
Ecosystems	Coral Springs FL	1-800-344-9003
Ecotek Inc	Fulton MO	1-800-242-3909
ENSCI Environmental Corp	High Point NC	1-800-241-9890
Enviro Spec Of Texas	Alvin TX	1-800-525-0520
Enviro-Land Services	Roseville CA	1-800-401-2225
Environmental Abatement Systems	Detroit MI	1-800-382-1531
Environmental Bio Control Intl	Elgin IL	1-800-468-6324
Environmental Compliance Services Inc	Las Vegas NV	1-800-327-1414
Environmental Exploration	Stockbridge GA	1-800-899-0475
Environmental Health & Safety Products	Prairie Village KS	1-800-779-3477
Environmental Institute For Tech Transf	Arlington TX	1-800-354-3350
Environmental Law Institute	Washington DC	1-800-433-5120
Environmental Management Intl	Englewood CO	1-800-352-7365
Environmental Media Corp	Chapel Hill NC	1-800-368-3382
Environmental Petroleum Service	Greenwood IN	1-800-821-7187
Environmental Products & Services Inc.	Springfield MA	1-800-545-4377
Environmental Recovery	Atlantic Beach FL	1-800-359-3740
Environmental Research Corps	E Freetown MA	1-800-899-2025
Environmental Specialties Inc	Sheboygan WI	1-800-246-7658
Environmental Technology	Claussen MI	1-800-229-3384
Environmental Training	Cinnaminson NJ	1-800-858-3358
Environmental Transport Corp	Ft Worth TX	1-800-886-3774
Eron Pest Control	Appleton WI	1-800-242-3552
Erosion Control Technologies	Branchburg NJ	1-800-245-0551
Eticam Of Rhode Island	Warwick RI	1-800-541-8673
Feecorp	Canal Winchester OH	1-800-828-9942
Fishery Information Management Systems	Auburn AL	1-800-659-8160
Florida Trail Assoc	Gainesville FL	1-800-343-1882
Flowmole Environmental Services	Kent WA	1-800-621-9374
Food & Water	Marshfield VT	1-800-328-7233
Fuelon Distributors	Coral Gables FL	1-800-321-5814
Future Horizons Inc	Hastings FL	1-800-682-1187
G & M Transportation	Rosedale MD	1-800-438-4216
Geoprojects Intl	Austin TX	1-800-553-7455
Gibraltar Chemical	Winona TX	1-800-256-5650
Global Perspectives	Sonoma CA	1-800-221-8897
GNI Group The	Deer Park TX	1-800-938-2526

Graymart Metal	Brooklyn NY	1-800-238-1813
Great Whales Foundation	San Francisco CA	1-800-421-9283
Ground Water Associate	Bridgewater NJ	1-800-331-0249
Guardian Environmental Services, Inc.	Bear DE	1-800-345-4395
Harmon Environmental Services Inc	Lafayette AL	1-800-835-9135
Harvard	Evansville WI	1-800-523-1327
Haz Tech Drilling	Meridian ID	1-800-359-1502
Healthy Homes	Brisbane CA	1-800-281-2515
Highland Energy Group	Boulder CO	1-800-736-8455
Hillman Environmental	Falls Church VA	1-800-882-4326
Hub Testing Laboratory	Waltham MA	1-800-878-8938
Hydro Turf & Associates	Gilberts IL	1-800-376-7466
In-Situ Inc	Laramie WY	1-800-446-7488
Indoor Air Quality Information Clearing	Washington DC	1-800-438-4318
Integrated Technical Services Inc.	Parsippany NJ	1-800-243-4320
Invisible Gardener-Organic Consultant	Malibu CA	1-800-354-9296
Jorge Scientific	Arlington VA	1-800-783-4862
Karen's Nontoxic Products	Conowingo MD	1-800-527-3674
Keep Florida Beautiful Inc	Tallahassee FL	1-800-828-9338
Kids For A Clean Environment	Nashville TN	1-800-952-3223
Kids For Saving Earth	Golden Valley MN	1-800-443-2784
Kil-All Pest Control	Antigo WI	1-800-704-9319
Kunafin	San Antonio TX	1-800-832-1113
L & H Consulting Inc	Ventura CA	1-800-500-5775
L A W E Enviormental Service	Newburyport MA	1-800-255-9265
Laidlaw Environmental Services	Columbia SC	1-800-356-8570
Lawco Inc	Epping NH	1-800-995-2926
Legato Trading Co	Kalispell MT	1-800-343-4445
Living Source The	Waco TX	1-800-662-8787
Lonestar Remediation Co.	Kountze TX	1-800-643-5250
Merrimack River Watershed Council	W Newbury MA	1-800-422-6792
Midwest Exterminating	Gobles MI	1-800-822-2366
National Container Network Inc.	Cleveland OH	1-800-345-0700
National Institute For Chemical Studies	Charleston WV	1-800-282-2796
Nature Spirit Products	Soquel CA	1-800-233-7309
Northwest Environmental Compliance Repo	Issaquah WA	1-800-642-8630
O H M Inc	Trenton NJ	1-800-562-2953
OHM Remediation Services Corp	Clermont FL	1-800-552-2038
On Site Environmental Service	Yankton SD	1-800-572-3272
Oppenheimer Environmental	Austin TX	1-800-835-2275
Palmetto Exterminators Inc	Walterboro SC	1-800-334-5122
Pedneault Associates Inc	Bohemia NY	1-800-356-8378
Pennsylvania Resources Council	Las Vegas NV	1-800-468-6772
Petroleum Wastewater Recycling	Oklahoma City OK	1-800-272-7971
Phillip Sams Pest Elimination	Bristol TN	1-800-633-8269
Plymouth Environmental	Bell PA	1-800-241-9920
Population Communications Intl	Newtown CT	1-800-322-8228
Priority Services Co	W Roxbury MA	1-800-559-5633
Protect Americas Eagles	Old Hickory TN	1-800-232-4537
R M I Environmental	Berlin NJ	1-800-220-8890
Radiation Technical Service Inc	Baton Rouge LA	1-800-336-3741
Resource Stratagies Inc	Port Leyden NY	1-800-895-7696
Rocky Mountain Elks Foundation	Missoula MT	1-800-225-5355
Ronnie Wall Inc	Diana TX	1-800-767-8475
RS Environmental Services Inc	Pompano Beach FL	1-800-940-6155
Save Our Streams	Glen Burnie MD	1-800-448-5826
Signal Environmental Services	Chattanooga TN	1-800-925-9551

SME Construction	Tacoma WA	1-800-613-9773
Soil Horizons Inc	Lafayette IN	1-800-288-7645
Soil Remediation	Englewood CO	1-800-441-1968
Solus System Inc.	Beaverton OR	1-800-247-5712
Sonag	Menomonee Falls WI	1-800-201-0990
Southern New England	Chepachet RI	1-800-772-8733
Spectrum Equip Intl	American Falls ID	1-800-455-2652
Sub Tech Inc	Evansville IN	1-800-448-4931
Superior Environmental	Romeoville IL	1-800-573-6847
Surfrider Foundation	San Clemente CA	1-800-743-7873
Synergic Analytics	Wyoming MI	1-800-446-5227
T P S Technologies Inc	Adelanto CA	1-800-862-8001
Tahoe Resource Conservation District	South Lake Tahoe CA	1-800-541-5654
Take It Back Foundation	Pacific Palisades CA	1-800-992-5389
Team	Loveland CO	1-800-782-7365
Team Marbo	Pearland TX	1-800-876-6476
TEC Environmental Laboratory	Jackson TN	1-800-832-1808
Techno Cycle Inc.	Hinesville GA	1-800-382-4491
Terra Environmental Systems	Dalton MN	1-800-782-9140
Texas Bio Remediation Council The	Richmond TX	1-800-266-9948
Thor Enterprises Southeast	Knoxville TN	1-800-282-8467
Timely Environmental Service Technology	Trenton NJ	1-800-837-8462
TPS Technologies Fax	Apopka FL	1-800-940-3222
Tri-State Environmental	Reedsville WV	1-800-362-2856
Unibest Inc	Bozeman MT	1-800-438-7734
United States Fish & Wildlife Service T	Annapolis MD	1-800-448-8322
Universal Recycled Water Systems Inc.	Orlando FL	1-800-532-6714
Valley Pest Control	Oshkosh WI	1-800-582-2847
Walden Walk A Thon	Boston MA	1-800-554-3569
Waste Management Inc	Phoenix AZ	1-800-672-0129
WEL Enterprises	Concord VA	1-800-847-2455
Wheelabrator Environmental Systems Inc	Hampton NH	1-800-682-0026
World Environmental Systems Ltd	Medford OR	1-800-525-1991
Wright R E Associate	Houston TX	1-800-545-9770

ENVIRONMENTAL, CONSERVATION & ECOLOGICAL ORGANIZATIONS

A 2000	Hohokus NJ	1-800-446-6050
Abasco	Houston TX	1-800-242-7745
Absolute Standards	Sheldon CT	1-800-368-1131
Air Chek Inc.	Arden NC	1-800-247-2435
Air Chek Inc.	Arden NC	1-800-257-2366
American Oceans Campaign	Santa Monica CA	1-800-862-3260
Animal Tracks	Eureka Springs AR	1-800-438-9949
Aqua Technologies	Plainview NY	1-800-635-9247
Barrier Solutions Inc.	Grants Pass OR	1-800-835-3607
Biological Mediation Systems	Ft Collins CO	1-800-524-1097
Carl Zent Co	Elko NV	1-800-345-8864
Les A. Cartier & Associates	Candia NH	1-800-537-4198
Chemserv Environmental Co.	Fairview OH	1-800-262-7744
Clean Environment Engineers Inc.	Emeryville CA	1-800-537-1767
Clean Environment Engineers Inc.	Norcross GA	1-800-441-5535
Clean Harbors Inc.	Baltimore MD	1-800-622-3360
Delaware River Keeper Network	Lambertville NJ	1-800-833-5292
Delta Institute	Austin CO	1-800-258-2921

Devore Enterprises	Sedona AZ	1-800-742-5553
Diversified Wastech Inc.	Levittown PA	1-800-762-2110
Dynamach	Rockville MD	1-800-346-6732
Earth Services Inc.	Ayden NC	1-800-221-0268
Earth Sites	Victoria TX	1-800-824-4097
Eco-Logica	Wayne NJ	1-800-525-2403
Ecosystems	Lauderdale Lakes FL	1-800-344-9003
Ecotek Inc.	Sturgeon MO	1-800-242-3909
Ecotopia	Greenbank WA	1-800-626-4242
Emissions Technology Inc.	Houston TX	1-800-382-7998
Endangered Species Federation	Northfield IL	1-800-732-8878
Energy Management Conservation Services		
	Woburn MA	1-800-982-4721
Ensco	Dalton GA	1-800-453-6726
Enviro Spec of Texas	Friendswood TX	1-800-525-0520
Environmental Audit Inc.	Lionville PA	1-800-542-8348
Environmental Bio Control International	Elgin IL (US ex DE)	1-800-468-6324
Environmental Defense Fund Hotline	Washington DC	1-800-225-5333
Environmental Management International	Englewood CO	1-800-352-7365
Environmental Remediation Systems	Waldorf MD	1-800-543-4347
Environmental Site Assessments	Mesa AZ	1-800-852-9512
Environmental Tech	Ardmore OK	1-800-634-4524
Environmental Technology of California	Thousand Oaks CA	1-800-628-1490
Environmental Training Inc.	Cinnaminson NJ	1-800-858-3358
Eticam of Rhode Island	Warwick RI	1-800-541-8673
Forest Resource Management Co.	Kirby VT	1-800-551-5213
Four Seasons Industrial Services	St Albans WV	1-800-626-8780
Great Whales Foundation	San Francisco CA	1-800-421-9283
Greenidea (Env educ)	Visalia CA	1-800-662-3427
Ground Water Associate	Bridgewater NJ	1-800-331-0249
Hadley Industries	Zelienople PA	1-800-872-9080
Innovate Waste Technologies	Kansas City MO	1-800-833-4340
Kuriger W	E Fitchburg MA	1-800-292-0921
L A W E Environmental Services	Newburyport MA	1-800-255-9265
Legato Trading Company	Virginia City MT	1-800-343-4445
Maritime Associates of the Port of New	New York NY	1-800-645-6772
McGown Reclamation Services	Mound City KS	1-800-223-9875
Meadows Company	Myrtle Beach SC	1-800-344-9001
Mid-Atlantic Environmental Inc.	Pasadena NC	1-800-245-7602
National Center for Toxicological Resea	Jefferson AR	1-800-638-3321
National Medical Advisory Service	Bethesda MD	1-800-258-0014
Nature Spirit Products	Silver Lakes WI	1-800-233-7309
The Northwest Environmental Compliance	Issaquah WA	1-800-642-8630
OHM Inc.	Princeton NJ	1-800-562-2953
Odis Envirotech	Plainview NY	1-800-553-9102
Oppenheimer Environmental	Austin TX	1-800-835-2275
Pennsylvania Dept Radiation Protection	Harrisburg PA	1-800-232-2786
Plymouth Environmental	Blue Bell PA	1-800-241-9920
Protect American Eagles	Old Hickory TN	1-800-232-4537
Re/Earth	Durant OK	1-800-542-1488
Refuse Industry Productions	Grass Valley CA	1-800-535-9547
Richardson Delaney	Lakewood NJ	1-800-545-6813
Rocky Mountain Elks Foundation	Missoula MT	1-800-225-5355
Soil Technologies	Charlotte NC	1-800-343-7645
TPS Technologies Inc.	Adelanto CA	1-800-862-8001
Tahoe Resource Conservation District	South Lake Tahoe CA	1-800-541-5654
Terra Environmental Systems	Dalton MN	1-800-782-9140

The Wolf & The Lamb	Ithaca NY	1-800-472-1169
Trees For The Future	New York NY	1-800-533-2784
Unibest Inc.	Boseman MT	1-800-438-7734
Venture II Environmental Drilling Inc.	Boerne TX	1-800-662-5412
Walden Walk A Thon	Boston MA	1-800-554-3569
Walden Woods Project	Boston MA	1-800-543-9911
World Environmental Systems Ltd.	Medford OR	1-800-525-1991
Wright RE Associate	Houston TX	1-800-545-9770

ENVIRONMENTAL CONSULTING FIRMS

3R Inc.	Greer SC	1-800-654-4434
A R S Technologies Inc	Ft Collins CO	1-800-346-5845
A1 Building Inspection Plus-Division Of	Capron IL	1-800-445-5290
AAA Environmental Services Inc	Ada OK	1-800-851-2519
AAA Environmental Services Inc.	Stratford OK	1-800-851-2519
AAAA By Phillips	Bristol PA	1-800-336-8813
Aaron Environmental Specialists	Waterbury CT	1-800-248-9858
Aaron Environmental Specialists	Waterbury CT	1-800-394-2374
ABB Environmental Services	Portland ME	1-800-341-0460
ABC Laboratories	Columbia MO	1-800-533-0222
Ablenet	Minneapolis MN	1-800-322-0956
Absolute Environmental Services	Westport CT	1-800-247-7906
ACC Environmental Consultants	Alameda CA	1-800-525-8838
Accoustical Systems Hearing Conservation Inc.	1-800-752-6640	
ACRT Inc. Environmental Specialists	Kent OH	1-800-622-2582
ACS Environmental	Conroe TX	1-800-793-1993
Acurex Environmental	Anaheim CA	1-800-800-2156
ADA Technologies Inc.	Englewood CO	1-800-232-0296
ADS Environmental Services Inc.	Indianapolis IN	1-800-554-1810
ADS Environmental Services Inc.	San Diego CA	1-800-533-3237
ADS Environmental Services Inc. (HQ)	Huntsville AL	1-800-633-7246
Advanced Analytical Solutions	Wayne PA	1-800-651-0878
Advanced Cleaning Systems/Dow Chemical		
	Midland MI	1-800-436-9227
Advanced Management Systems	Corpus Christ TX	1-800-453-9923
Advanced Tank Technology	Ft Worth TX	1-800-526-1446
Aegis Environmental Management	Milford OH	1-800-328-6878
Aegis Environmental Management	Milford OH	1-800-331-0269
AETC	Marlborough MA	1-800-354-2382
Agassiz Environmental	Cedar MN	1-800-362-7087
Agency Information Consultants	Austin TX	1-800-945-9509
AgriGrowth Research Inc.	Indianapolis IN	1-800-331-5179
AgriManagement Inc.	Yakima WA	1-800-735-6368
AIC Research Inc	Washington DC	1-800-328-3390
AIC Research Inc.		1-800-328-3390
Ainley Envirotech	Algonquin IL	1-800-962-8694
Air Science Consultants	Bridgeville PA	1-800-759-9282
Alden Analytical Laboratories Inc.	Seattle WA	1-800-669-3660
All State Engineering & Testing Consult	Miami FL	1-800-826-6857
All States Environmental Services Inc.	St Charles MO	1-800-851-5446
Alliance Environmental Inc.	Marietta OH	1-800-322-6653
Allied Signal Environmental Systems &	SMorristown NJ	1-800-626-4974
Allstate Power Vac	Huntington NY	1-800-331-7733
Allstate Power Vac	Huntington NY	1-800-331-7753
Allstate-Nevada Environmental Mgt	Grs Valley NV	1-800-851-6527
Allstate-Nevada Environmental Mgt	Winnemuca NV	1-800-851-6527

Allwaste Environmental Services	Golden CO	1-800-448-1799
American Digital System	Huntsville AL	1-800-633-7246
American Environmental Engineering	Leeds AL	1-800-238-8744
American Environmental Safe Home	New Bloomfield PA	1-800-437-2749
American Environmental Technology	De Soto MO	1-800-328-2382
American Eviromental Mgmt Corp	Rancho Cordova CA	1-800-332-2362
American Fluids International	Conroe TX	1-800-845-3517
American Hazard Control Consultants Inc	Caldwell NJ	1-800-326-4051
American Laboratory For Environmental	E Frankfort IL	1-800-533-2539
American Laboratory for Environmental	EFrankfort IL	1-800-533-2539
American Oil & Industrial Services Inc.	Covington GA	1-800-322-2197
American Radiation Services Inc	Baton Rouge LA	1-800-401-4277
American Services Associates	Issaquah WA	1-800-869-7784
American Standards Testing Bureau Inc.	New York NY	1-800-221-5170
American Waste Processing Ltd.	Maywood IL	1-800-841-6900
American Water Works Association	Denver CO	1-800-926-7337
Ameriwaste Environmental Inc.	Cleveland OH	1-800-343-2179
Analytica Inc.	Golden CO	1-800-873-8707
Analytical Environmental Services	Atlanta GA	1-800-972-4889
Analytikem Inc.	Cherry Hill NJ	1-800-879-5221
Andersen 2000	Peachtree City GA	1-800-241-5424
Antimicrobial Products Complaint System	Lubbock TX	1-800-447-6349
Antimicrobial Products Compliant System	Lubbock TX	1-800-447-6349
APEC Ltd.	Kalamazoo MI	1-800-PIK-APEC
APEC Ltd.	Kalamazoo MI	1-800-745-2732
Apollo Environmental Strategies	Beaumont TX	1-800-742-1033
Applied Environmental Science Inc.	Minneapolis MN	1-800-626-8089
Applied Respiratory Technology		1-800-966-4330
Applied Respiratory Technology (ART)	Cold Spring NY	1-800-966-4330
Applied Science & Technology Inc. (ASTI)	Ann Arbor MI	1-800-395-ASTI
Aptus Inc.	Lakeville MN	1-800-328-4061
Aqua Sierra Fisheries Consultants Inc.		1-800-524-3474
Aqua Survey	Flemington NJ	1-800-654-4684
Aqua Survey	Flemington NJ	1-800-654-4684
Aqua Tech Environmental Consultants Inc	Marion OH	1-800-783-5991
Aqua Tech Environmental Consultants Inc	Sanford NC	1-800-522-2832
Aqua Tech Environmental Consultants Inc	Canton OH	1-800-635-3222
Aqua Tech Environmental Consultants Inc	Melmore OH	1-800-858-8869
Aqua-Flo Inc.	Baltimore MD	1-800-368-2513
Aquarian Management Services	Monroe CT	1-800-832-3794
Aquatic Conservation Inc.	St Petersburg FL	1-800-521-4075
Archimedes Environmental Equipment Inc	.Beaumont TX	1-800-582-4187
ARI Network Services Inc.	Milwaukee WI	1-800-558-9044
Aries Group	Austin TX	1-800-222-1685
Arizona Testing Laboratories	Phoenix AZ	1-800-279-6181
ARS Technologies Inc.	Ft Collins CO	1-800-346-5845
Asbestos Research & Environmental Assoc	Congers NY	1-800-245-2732
Asbestos Safety Technologies Of Califor	Santee CA	1-800-742-5365
Asbestos Safety Technologies of Califor	Santee CA	1-800-742-5365
Ascheman Associates Consulting (Agri)	Des Moines IA	1-800-798-7371
Associated Environmental Consultants	Raleigh NC	1-800-447-3235
ASTB/Analytical Services Inc.	New Castle DE	1-800-221-5170
ATC Environmental Inc.	Los Angeles CA	1-800-227-6935
ATC Environmental Inc.	Sioux Falls SD	1-800-522-9675
ATEC Associates Inc.	Indianapolis IN	1-800-473-0194
Atlantic Environmental	Dover NJ	1-800-344-4414
Atlantic ESI	Delmar NY	1-800-221-2881
Atlantic Petroleum Tank Service	Manchester NH	1-800-882-6587

Auburn Environmental Consulting & Testi	Auburn IL	1-800-662-1584
Auchter Industrial Vac Of Florida	Melbourne FL	1-800-258-4377
Axact Standards Inc	Amityville NY	1-800-528-6502
Axact Standards Inc.	Commack NY	1-800-528-6502
Azimuth Inc.	Charleston SC	1-800-537-0336
Aztec Environmental Services	Norman OK	1-800-879-0255
B & A Engineers Inc.	Boise ID	1-800-327-5360
B & V Waste Science & Technology Corp.	Kansas City MO	1-800-528-9278
Badger Laboratories & Engineering Co.	Appleton WI	1-800-776-7196
Baker Environmental Inc.	Coraopolis PA	1-800-553-1153
Baker Environmental Inc.	Merrillville IN	1-800-473-6847
Baker Environmental Inc.	Westboro MA	1-800-334-8011
Baker Pacific Corp	Long Beach CA	1-800-346-3809
Balsam Environmental Consultants Inc.	Salem NH	1-800-933-9322
Basic Environment Group Inc	Linn Creek MO	1-800-437-7993
Batta Environmental Associates	Newark DE	1-800-543-4807
Bay West Inc.	St. Paul MN	1-800-279-0456
BD & S Computer Services		1-800-323-2765
Beede Waste Oil	Plaistow NH	1-800-562-9198
Benchmark Engineering Inc.	Auburn IL	1-800-325-0011
Berghoff/America Inc.	Concord CA	1-800-544-5004
Beta Environmental Engineers	Duncanville TX	1-800-887-8811
BHA Group Inc.	Kansas City MO	1-800-821-2222
Biggins W F Associates Inc	Meadow MA	1-800-722-3563
Bio Clean Environmental Inc.	Pocasset MA	1-800-339-9728
Bio-Concepts Inc	Seabrook TX	1-800-828-5124
Bio-Concepts Inc.	Kemah TX	1-800-828-5124
Bio-Genesis Inc.	Framingham MA	1-800-458-1155
Bio-Organic Composting Products	Little Rock AR	1-800-982-3950
Bio-Quest-Bacterial Products & Services	Bakersfield CA	1-800-645-2847
Bio-Rem	Butler IN	1-800-428-4626
Bio-Technics Laboratories Inc.	Los Angeles CA	1-800-331-9958
Biologic Consultants Ltd.	Allentown PA	1-800-458-9515
Bioscience Inc.	Bethlehem PA	1-800-627-3069
Biosystems/Atlanta Div. Biosystems Inc.	Loganville GA	1-800-346-2467
Biotech Development Corp.	Marietta GA	1-800-842-6117
Biotransformation Remedial Technologies	Colorado Springs CO	1-800-642-8926
Black W L and Associates	Chesapeake VA	1-800-669-7925
Blair Tree Experts	Big Pool MD	1-800-262-0800
Blando Environmental	Anderson SC	1-800-257-2287
Blymer Engineers Inc.	Alameda CA	1-800-753-3773
Bodine Environmental Services	Decatur IL	1-800-637-2379
Boermans Environmental Specialties & Tr	Allegan MI	1-800-442-3781
Brevard Oil Equipment	Melbourne FL	1-800-472-7670
Brooks Companies	Weston CT	1-800-843-1631
Brown & Root Environmental	Houston TX	1-800-368-2755
Brown & Root Environmental	Gaithersburg MD	1-800-368-2755
Brown and Caldwell Consultants	Pleasant Hill CA	1-800-727-2224
Bryson Industrial Services	Memphis TN	1-800-822-8146
BSC Group	Boston MA	1-800-288-8123
Buffalo Environmental Products Corp	Baltimore MD	1-800-462-3526
Burlington Environmental	Seattle WA	1-800-228-7872
C C Enviroklean	Kansas City KS	1-800-643-7049
C F C I Inc	Lipan TX	1-800-453-4510
C G R S	Ft Collins CO	1-800-288-2657
CAE Diagnostic Services	Palatine IL	1-800-627-3300
CAE Diagnostic Services	Palatine IL	1-800-627-0033
California Business Environmental	Anaheim CA	1-800-662-2322

California State Health Service Toxic	SSacramento CA	1-800-334-1697
Camp Dresser & McGee Inc.	Cambridge MA	1-800-243-2677
Camp Dresser & McGee Inc.	Cambridge MA	1-800-343-7004
Camp Dresser & McKee	Cambridge MA	1-800-243-2677
Can Do Service Co	Melbourne Beach FL	1-800-828-2820
Capsule Environmental Engineering	Roseville MN	1-800-328-8246
Capsule Environmental Engineering Inc.	St. Paul Mn	1-800-328-8246
Carbonair Inc.	Maple Grove MN	1-800-526-4999
Carbtrol® Corp	Westport CT	1-800-242-1150
Carnow Conibear & Associates	Rockville MD	1-800-562-0419
Caron Products & Services Inc.	Marietta OH	1-800-648-3042
Carylon Corporation	Chicago IL	1-800-621-4342
CBE Inc	Boca Raton FL	1-800-782-0411
CBE Inc.	Boca Raton FL	1-800-782-0411
CC Enviroklean	Kansas City KS	1-800-643-7049
CCI Technologies	San Francisco CA	1-800-752-1415
Center for Holistic Resource Management	Albuquerque NM	1-800-654-3619
Central New York Industrial Service Inc	Greensboro NC	1-800-262-9374
Central States Environmental Services I	Centralia IL	1-800-367-3090
Certified Engineering and Testing	Weymouth MA	1-800-443-4353
Certified Environmental Consulting	Murray UT	1-800-231-2992
Certified Environmental Consulting	Roseville CA	1-800-453-3136
Certified Environmental Consulting	Salt Lake City UT	1-800-231-2992
Certified Environmental Consulting	Benecia CA	1-800-228-0171
Certified Environmental Consulting	San Ramon CA	1-800-447-0171
Certified Environmental Consulting	Phoenix CA	1-800-554-0288
CET Engineering Services	Huntington PA	1-800-643-8260
CET Environmental Services Inc.	Long Beach CA	1-800-753-1818
Chama Environmental Drilling Systems	Roanoke IN	1-800-232-4262
Chematox Laboratory Inc. (forensics)	Boulder CO	1-800-334-1685
Chemical Safety & Compliance	Southfield MI	1-800-451-5498
Chemical Waste Management	N. Palm Beach FL	1-800-451-5498
Chemical Waste Management Inc.	Oak Brook IL	1-800-843-3604
Chemical Waste Management Inc.	Henderson CO	1-800-525-1840
Chemical Waste Management Inc.	Tukwila WA	1-800-843-3604
Chempet Research Corp	Moorpark CA	1-800-473-1085
Chesapeake Software	Chadds Ford PA	1-800-229-5813
Cistar Technologies Inc	Canton MA	1-800-992-9600
City Environmental Contracting Inc.	Detroit MI	1-800-992-9118
Civil & Environmental Consultants Inc.	Pittsbugh PA	1-800-365-2324
Civil & Environmental Consultants Inc.	Cincinnati OH	1-800-759-5614
CKY Inc.	Torrance CA	1-800-388-8889
Clarus Systems Group	Laguna Miguel CA	1-800-223-1998
Clay, Sovich Environmental Inc.	Manassas VA	1-800-348-7427
Clean Air Engineering	Palatine IL	1-800-627-0033
Clean Environment Engineers	Norcross GA	1-800-441-5535
Clean Environment Equipment	Oakland CA	1-800-537-1767
Clean Harbors Inc.	Farmington CT	1-800-637-2666
Clemen Environmental Services	Wilton CA	1-800-352-4432
Coastal Remediation Co.	Roanoke VA	1-800-776-5733
Cochrane Associates	Wheat Ridge CO	1-800-635-9582
Cognis Inc	Santa Rosa CA	1-800-524-3307
Cognis Inc.	Santa Rosa CA	1-800-524-3307
Compacting Technologies International	Portland OR	1-800-727-2067
Compliance Systems Inc.	Bellmawr NJ	1-800-622-2527
Comprehensive Environmental	Newport Beach CA	1-800-292-7834
Comprehensive Environmental	Newport Beach CA	1-800-292-7834
Comprehensive Loss Management	Minneapolis MN	1-800-533-2767

Con Test	Williston VT	1-800-331-5967
Con-Test Labs Inc.	E Longmeadow MA	1-800-634-8165
Concord Environmental Services Inc.	Cuyahoga Falls OH	1-800-252-3098
Confidential Compliance Consultants	San Jose CA	1-800-833-9938
Connecticut Test Borings	Seymour CT	1-800-782-8085
Conservation Services Inc.	San Antonio TX	1-800-452-2049
Consolidated Environmental Services	Perrysburg OH	1-800-777-0782
Consolve	Lexington MA	1-800-241-2431
Consultants & Contractors	Chantilly VA	1-800-322-3477
Consulting Services Inc.	Exton PA	1-800-858-0853
Container Testing Laboratory	Mamaroneck NY	1-800-221-5170
Contest Environmental Education Center	E Longview MA	1-800-626-8378
Continental Environmental Services	Gainesville FL	1-800-342-1103
Continental Weather Services Inc.	Encino CA	1-800-843-7246
Controls for Environmental Pollution In	Sante Fe NM	1-800-545-2188
Cornerstone Global Environmental Techno	Waco TX	1-800-432-0603
Cortest Laboratories Inc.	Houston TX	1-800-352-2254
CPI Chicago Inc.	Lake Forest IL	1-800-737-LABS
Crane Institute of America Inc.	Maitland FL	1-800-832-2726
Crawford & Company (Risk Control)	Atlanta GA	1-800-241-2541
Crosby & Overton	Oakland CA	1-800-821-0424
Crouse Enterprises	Pittsburgh PA	1-800-927-6873
CSI	Exton PA	1-800-858-0853
CTC-Geotek Inc.	Salt Lake City UT	1-800-843-6835
CTI Environmental Services	Toledo OH	1-800-828-9096
Cummings & Smith Inc	Hershey PA	1-800-533-8868
CYN Environmental Services	Stoughton MA	1-800-622-6365
D-Tox Environmental Contractors	Newington CT	1-800-336DTOX
Dalsin Industries	Columbus OH	1-800-442-4849
Dames & Moore	Farmington Hills MI	1-800-962-1177
Daniel J. Hartwig Associates Inc.	Oregon WI	1-800-837-4411
Darell Bevis Associates Inc.	Chantilly VA	1-800-368-0335
DAS International Inc.	Exton PA	1-800-966-2420
David Cornish & Co.	Ft Worth TX	1-800-621-5993
Dawkins Arnold	Kingsland TX	1-800-278-5788
Delta Environmental Consultants	Rancho Cordova CA	1-800-521-0786
Delta Environmental Consultants	St. Paul MN	1-800-477-7411
Delta Environmental Consultants	Arden Hills MN	1-800-888-1331
Delta Environmental Consultants		1-800-521-0786
Delta Environmental Consultants	Tampa FL	1-800-223-5951
Diamond Environmental Services Ltd.	Acampo CA	1-800-847-4793
Digicolor Inc.	Columbus OH	1-800-848-6448
Digicolor Priority Instrument Rental	Columbus OH	1-800-848-6448
Discovery Recycling Consultants Inc	Lanoka Harbor NJ	1-800-662-5040
Discovery Recycling Consultants Inc.	Lanoka Harbor NJ	1-800-662-5040
DJP Asswociates Ltd.	Cheshire CT	1-800-262-0506
DNA Industrial Hygiene	Lawndale CA	1-800-644-1924
Down Under Tank Testing Inc.	N Palm Beach FL	1-800-782-8265
Downey Associates Inc.	Ft Valley GA	1-800-528-3046
Dredge Masters	Toms River NJ	1-800-237-4021
Dupont Safety & Environmental Mgt	Wilmington DE	1-800-532-SAFE
Dupont Safety & Environmental Mgt	Wilmington DE	1-800-532-7233
Dustvent Inc.	Addison IL	1-800-553-3687
E T C	Birmingham AL	1-800-677-8761
EA Engineering, Science & Technology	Charlotte NC	1-800-825-9097
Eagle-Picher Environmental Services	Miami OK	1-800-331-7425
Earl Ruble & Associates Inc.		1-800EHRUBLE
Earth & Atmospheric Sciences Inc.	Dayton OH	1-800-543-9930

Earth Guard Environmental Inc	Henderson WV	1-800-559-0291
Earth Guard Environmental Inc.	Las Vegas NV	1-800-554-0291
Earth Protection Systems Inc.	Bridgeview IL	1-800-553-4377
Earth Science Technology	Northampton MA	1-800-826-5006
Earth Systems Consultants	Palo Alto CA	1-800-523-9341
Earyth Resources Corp.	Ocoee FL	1-800-877-0850
Eastern Analytical Services Inc	Elmsford NY	1-800-327-0121
Ebasco Environmental	Lyndhurst NJ	1-800-580-3765
EBN Inc.	Muskegon MI	1-800-475-3291
ECAR	Virginia Beach VA	1-800-831-0913
EcoChem Technologies Inc.	West Hills CA	1-800-340-3090
Ecotec	Fraser CO	1-800-932-6832
ECS Inc.	Exton PA	1-800-327-1414
ECT Environmental Control Technologies		1-800-331-1945
ECT Environmental Control Technologies	Bedford NH	1-800-962-3755
Eddie Taw Inc.	Mesquite TX	1-800-922-2051
EDR/Toxicheck(Environmental Records Sea	Fairfield CT	1-800-352-0050
EHMI	Durhamville TN	1-800-558-3464
EIS International	Rockville MD	1-800-999-5009
EMAC Inc.	Pittsburgh PA	1-800-533-4187
EMI	Elburn IL	1-800-252-7000
Empire Environmental Inc.	Stuart FL	1-800-344-1087
EMR/Environmental Management Res	Bedford NH	1-800-688-0979
Emulsions Control Inc.	National City CA	1-800-433-6857
Enasco Inc.	Buffalo NY	1-800-851-4482
Encapsulation Technologies	Baltimore MD	1-800-942-0145
Endispute/ADR	Boston MA	1-800-892-0239
Enechotech	Denver CO	1-800-814-9532
Eneco Tech Inc.	Denver CO	1-800-659-TECH
Enercon Services Inc.	Tulsa OK	1-800-735-7693
Energy Alliance Inc.	Cincinnati OH	1-800-735-0359
Engineering Design & Testing Corp.	Columbia SC	1-800-338-5227
ENSR Consulting & Engineering	Acton MA	1-800-722-2440
ENSR Health Sciences	Alameda CA	1-800-922-4636
ENSR Operations	Canton OH	1-800-759-3677
Entek Environmental Corp	Lagrange IN	1-800-551-2520
Entropy Environmentalists Inc.	Research Triangle Park NC	1-800-486-3550
Enviro Chem		1-800-247-9011
Enviro Group Inc.		1-800-368-4761
Enviro Med II		1-800-582-8689
Enviro Physics		1-800-228-2257
Enviro Probe	Bronx NY	1-800-833-3585
Enviro Probe	Edison NJ	1-800-833-3585
Enviro Supply Co.	Palatine IL	1-800-992-8969
Enviro Tech Mid Atlantic	Blackburg VA	1-800-828-3862
Enviro Tech Service	Pittsburg CA	1-800-468-8921
Enviro Tech Services	Martinez CA	1-800-468-8941
Enviro-Med Laboratories	Ruston LA	1-800-256-4362
Enviro-Zyme	Stormville NY	1-800-882-9904
Envirochem		1-800-346-4876
Envirocorp Services & Technology Inc.	Houston TX	1-800-535-4105
Envirocraft Corp.	Bellmawr NJ	1-800-223-4525
Envirolytics Inc	Sarasota FL	1-800-854-9456
Enviromark	Davenport IA	1-800-368-3662
Enviromed		1-800-822-5800
Enviromed	Raleigh NC	1-800-554-8387
Enviromed Systems		1-800-468-7493
Environ	Princeton NJ	1-800-552-3272

Environ	Princeton NJ	1-800-344-2986
Environamics Corp.		1-800-241-1666
Environetics Inc.		1-800-637-0815
Environmental & Regulatory Consultants	Raleigh NC	1-800-446-4372
Environmental Abatement Support		1-800-548-2893
Environmental Abatement Systems	Detroit MI	1-800-382-1531
Environmental Appraisers The	Corpus Christi TX	1-800-551-2532
Environmental Assessment Inc	W Paterson NJ	1-800-580-4242
Environmental Audit Inc.	Lionville PA	1-800-542-8348
Environmental Business Journal		1-800-446-4325
Environmental Chemical Co.		1-800-262-0458
	Salt Lake City UT	1-800-362-4369
Environmental Cleaning Concepts	Loveland OH	1-800-447-1322
Environmental Compliance Consulting	Matthews NC	1-800-723-2023
Environmental Compliance Consulting Ser	Tulsa OK	1-800-972-6420
Environmental Compliance Consulting Ser	Iowa City IA	1-800-972-6420
Environmental Compliance Services Inc.	Exton PA	1-800-327-1414
Environmental Compliance Systems Inc	Lake Forest IL	1-800-472-8882
Environmental Compliance Systems Inc.	Lake Forest IL	1-800-472-8882
Environmental Considerations Ltd. Arizona		1-800-255-6530
Environmental Construction Outfitters	New York NY	1-800-238-5008
Environmental Construction Services	Castleton On Hudson NY	1-800-824-5153
Environmental Consulting Services Inc.	Lake Worth FL	1-800-585-6973
Environmental Contracting	Alsip IL	1-800-331-1945
Environmental Control Division	Memphis TN	1-800-237-9030
Environmental Control Experts		1-800-821-3022
Environmental Control Technologies	Nashua NH	1-800-962-3755
Environmental Crisis Management Inc.	Stanford CT	1-800-338-4490
Environmental Data Resources	Fairfield CT	1-800-352-0050
Environmental Database	Littleton CO	1-800-982-4627
Environmental Design Associates		1-800-247-3658
Environmental Diagnostics		1-800-334-1116
Environmental Dynamics Inc.	Fredericksburg VA	1-800-448-4723
Environmental education Enterprises Inc	Columbus OH	1-800-792-0005
Environmental Engineering Construction	Conyers GA	1-800-358-4816
Environmental Engineering Lab		1-800-231-4669
Environmental Enterprise of Florida Inc	Orlando FL	1-800-223-8140
Environmental Enterprise of Florida Inc	Ellenwood GA	1-800-238-3230
Environmental Enterprises	Cincinnati OH	1-800-392-1503
Environmental Enterprises Of Florida	Ellenwood GA	1-800-238-3230
Environmental Evaluation	Toms River NJ	1-800-852-2702
Environmental Hazard Control		1-800-543-1499
Environmental Hazards Consulting Inc.	Lancaster PA	1-800-338-3424
Environmental Health & Light Research I	Tampa FL	1-800-544-4878
Environmental Health Associates		1-800-922-4636
Environmental Health Associates		1-800-342-4636
Environmental Hygiene Associates	Boothwyn PA	1-800-442-4342
Environmental Impex International		1-800-221-7458
Environmental Information Systems		1-800-541-4190
Environmental Innovations		1-800-554-0037
Environmental Institute for Technology Transfer		1-800-354-3350
Environmental Instruments	Concord CA	1-800-648-9355
Environmental Instruments	Concord CA	1-800-453-4526
Environmental IQ		1-800-992-9754
Environmental Law Associates		1-800-443-0699
Environmental Logistics		1-800-227-2850
Environmental Maintenance Solutions		1-800-962-3310
Environmental Management	St. Clair Shores MI	1-800-531-5576

Environmental Management & Compliance	Little Falls NJ	1-800-345-4362
Environmental Management & Design	Reading PA	1-800-883-7956
Environmental Management & Planning	Chico CA	1-800-362-2921
Environmental Management Consultants In	Phoenix AZ	1-800-362-3373
Environmental Management Group Inc.	Shoreham NY	1-800-343-3644
Environmental Management Inc.	Atlanta GA	1-800-542-5244
Environmental Management International	Englewood CO	1-800-352-7365
Environmental Management International	Englewood CO	1-800-352-7365
Environmental Management Systems Inc.	Altamonte Springs FL	1-800-654-5176
Environmental Manufacturing & Supply	Bonifay FL	1-800-624-0005
Environmental Medicine Inc.	Westwood NJ	1-800-468-6942
Environmental Mfg & Supply		1-800-624-0005
Environmental Monitor Service Inc.	Yalesville CT	1-800-864-2814
Environmental Monitoring Contractors	Gilberts IL	1-800-331-2925
Environmental Monitoring Contractors	Gilberts IL	1-800-331-2925
Environmental Monitoring Corp	Hartwell GA	1-800-848-7437
Environmental Oil Inc.	Syracuse NY	1-800-262-1012
Environmental Options		1-800-346-4606
Environmental Plus	Evergreen CO	1-800-368-7587
Environmental Protection Inc.	Homewood IL	1-800-526-1788
Environmental Protection Inc.	Kalkaska MI	1-800-221-2535
Environmental Protection Inc.	Mancelona MI	1-800-345-4637
Environmental Protection Inspection & Consulting		1-800-292-7635
Environmental Protection Inspection & C	Kansas City MO	1-800-421-8302
Environmental Protection Inspection and Consulting		1-800-421-8302
Environmental Public Relations Inc.		1-800-346-4288
Environmental Quality Laboratories		1-800-368-5227
Environmental Realty Guild of America	Philadelphia PA	1-800-462-3742
Environmental Reclamation Systems	Port St. Lucie FL	1-800-457-0222
Environmental Rental Inc		1-800-446-8736
Environmental Research Consultants Inc.		1-800-533-7858
Environmental Resource Associates	Warwick RI	1-800-338-7398
Environmental Resource Associates	Arvada CO	1-800-372-0122
Environmental Resource Associates of Florida Inc.		1-800-221-3486
Environmental Resource Center	Cary NC	1-800-537-2372
Environmental Response Inc.	Hendersonville TN	1-800-582-5148
Environmental Restoration Co	Fredericksburg VA	1-800-570-1757
Environmental Restoration Group	Palm City FL	1-800-454-7347
Environmental Risk Information and Imag	Alexandria VA	1-800-989-0402
Environmental Safeguards		1-800-222-9265
Environmental Science & Engineering	Gainseville FL	1-800-874-7872
Environmental Science & Engineering	Peoria IL	1-800-ESE-1999
Environmental Science Lab		1-800-343-0707
Environmental Service & Technology		1-800-562-2118
Environmental Service Company	Chamblee GA	1-800-533-3273
Environmental Service Drilling		1-800-845-8470
Environmental Service Group of New York	Tonnawanda NY	1-800-348-0316
Environmental Services Group	Englewood Cliffs NJ	1-800-877-2436
Environmental Services Inc.	Jacksonville FL	1-800-443-3158
Environmental Site Assessments		1-800-852-9512
Environmental Solutions Group	ryville IN	1-800-392-5474
Environmental Source Samplers	Huntersville NC	1-800-245-3778
Environmental Specialty Inc.		1-800-451-6096
Environmental Systems		1-800-732-2002
Environmental Systems & Technologies	Blacksburg VA	1-800-926-5923
Environmental Systems and Service	Farmington CT	1-800-446-8424
Environmental Tech Intl	San Jose CA	1-800-244-7202
Environmental Technical Services		1-800-634-2103

Environmental Technologies Group Inc.	Baltimore MD	1-800-635-4598
Environmental Technologies Inc	Richmond VA	1-800-462-4046
Environmental Technologies Inc.	Magnolia TX	1-800-683-1046
Environmental Technology	Bowling Green OH	1-800-866-0022
Environmental Technology of North Ameri	Richmond VA	1-800-228-7746
Environmental Technoloy Inc	Richmond VA	1-800-228-7745
Environmental Test System Inc.		1-800-446-2014
		1-800-548-4381
Environmental Testing Center	Lighthouse Point FL	1-800-272-8966
Environmental Toxicology International	Seattle WA	1-800-669-4384
Environmental Training Consultants Inc.		1-800-451-7747
Environmental Treatment & Technologies	Findlay OH	1-800-822-3882
Environmental Waste Management Inc.	Lees Summit MO	1-800-533-2846
Environmental Waste Resources Inc.		1-800-225-5397
Environmental Waterway Management	Plantation FL	1-800-832-5253
Environments Inc.	Beaufort SC	1-800-342-4453
Environmetrics Inc.	Maryland Heights MO	1-800-333-3278
Enviropore Inc.	Lumberton NJ	1-800-874-6270
EnviroPro Inc.	Chatsworth CA	1-800-533-8378
Enviroscan Corp	Rothschild WI	1-800-338-7226
Envirosurv	Arlington VA	1-800-243-3580
Envirotec Operating Services Inc.		1-800-331-8482
Envirotech	Farmington NM	1-800-362-1879
Envirotech Systems	Seattle WA	1-800-922-9395
Envirowaste Management Institute of Ame	Kansas City MO	1-800-527-9537
Envisage Environmental Inc.	Richfield OH	1-800-878-0990
EPCON Industrial Systems Inc.	The Woodlands TX	1-800-447-7872
EPG Companies Inc.	Maple Grove MN	1-800-443-7426
Epic Training/Environmental Services In	Kansas City MO	1-800-421-8302
ERAtech Environmental Inc.	Dayton OH	1-800-848-4990
ERAtech Inc.	Dayton OH	1-800-365-2146
EssTek Inc.	Englewood CO	1-800-753-8504
Etc./Pacific Northwest Environmental La	Redmond WA	1-800-292-7635
ETEC	Los Lunas NM	1-800-831-1085
ETEC Remediation Service	Belen NM	1-800-831-1085
Evergreen Analytical Inc.	Wheat Ridge CO	1-800-845-7400
Evergreen Services Corp	Bellevue WA	1-800-726-1751
Exploration Resources	Athens GA	1-800-231-3282
Feehan Corp	South Bend IN	1-800-392-8411
Ferguson Harbour Inc.		1-800-822-3295
Fertech Enviro Systems Inc.	Moberly MO	1-800-362-8808
Firecon	East Earl PA	1-800-222-8841
First Environment Inc.		1-800-486-5869
Fitt Environmental	Ridgefield WA	1-800-224-3488
Florida Environmental Engineering Inc.	Pompano Beech FL	1-800-845-7816
Florida Waste & Environmental	Odessa FL	1-800-554-8476
Florida Waterway Management	Plantation FL	1-800-832-5253
Forecon Inc. (Forestry)	Jamestown NY	1-800-527-7799
Forest Consulting Service Inc.	Pleasant Hill OR	1-800-869-3915
Foth & Van Dyke Engineering	Green Bay WI	1-800-836-2500
Friendly Water & Air	Dewitt MI	1-800-457-2825
Funderburk & Associates		1-800-227-6543
Future Horizons Inc.	Hastings FL	1-800-682-1187
G C I Inc	Farmingdale NY	1-800-842-5073
Galson Corp.	E Syracuse NY	1-800-950-0506
Gandee & Associates Inc.	Columbus OH	1-800-442-4849
Gann Associates	Glen Elyn IL	1-800-762GANN
Gannett Fleming Engineers Inc.	Harrisburg PA	1-800-233-1055

Garrow and Associates	Atlanta GA	1-800-868-2948
Gas Monitoring	Cary NC	1-800-438-1152
GEI Consultants Inc.	Winchester MA	1-800-678-1501
Gelbert & Company Forestry Consultants	Durham NC	1-800-277-3326
Gem Star Associates Inc.	Waubay SD	1-800-456-3305
Geo-Technique	Southlake TX	1-800-348-6308
Geo-Technique	Southlake TX	1-800-828-1407
Geologic Exploration	Statesville NC	1-800-752-8853
Geomonitoring Services	Houston TX	1-800-373-0808
Geonetics Corp	Boone NC	1-800-425-7117
Geoquip Inc	North Plains OR	1-800-848-1212
Geoquip Inc.	Hillsboro OR	1-800-848-1212
Georesearch	Long Beach CA	1-800-523-4786
George Analytics GAI	Nicholasville KY	1-800-554-8006
Geotechnical Service North Inc.	Dublin CA	1-800-643-6832
Geotest	Long Beach CA	1-800-624-5744
Geraughty & Miller	Bellevue WA	1-800-225-8419
Geraughty & Miller	Plainview NY	1-800-225-8419
Gilarde Environmental Management	Dunmore PA	1-800-356-2623
Gillard Environmental Management	Dickson City PA	1-800-356-2623
Gilson Co. Inc.	Worthington OH	1-800-444-1508
Global Environmental Inc.	San Diego CA	1-800-437-7873
Global Environmental Inc.	San Diego CA	1-800-462-9283
Global Environmental Technologies Inc.	Allentown PA	1-800-800-8377
Global Spill Management Inc.	Valley Forge PA	1-800-258-5585
Goldman Environmental Consultants Inc.	Braintree MA	1-800-446-2014
Graham & Curie Well Drilling Co. Inc.	West End NC	1-800-368-6628
Grease Busters Service	Braintree MA	1-800-342-1187
Great Lakes Carbon Treatment	Kalkaska MI	1-800-841-8324
Great Lakes Spill Co-Op	Detroit MI	1-800-425-6557
Greeley and Hansen	Chicago IL	1-800-837-9779
Greeley and Hansen	Chicago IL	1-800-824-7932
Ground Water & Environmental Services	E Providence RI	1-800-680-0414
Ground Water & Environmental Services	Lionville PA	1-800-426-9871
Groundwater & Environmental Services In	Wall NJ	1-800-220-3068
Groundwater & Environmental Services In	Stamford CT	1-800-358-9114
Groundwater Recovery Systems	Concord CA	1-800-221-3795
Groundwater Services International Inc.	Middletown PA	1-800-528-4289
Groundwater Technology Inc	Norwood MA	1-800-635-0053
Groundwater Technology Inc.	Norwood MA	1-800-635-0053
GSX Laidlaw Environmental Services	Taft CA	1-800-752-6867
GT Environmental Labs	Concord CA	1-800-544-3422
GTEL Environmental Lab Inc.	Carollton TX	1-800-235-0977
GTS Duratek	Columbia MD	1-800-638-3838
GTS Duratek	Columbia MO	1-800-338-4487
Gulf Environmental Inc	La Porte TX	1-800-448-0209
H & H Toxic Technology	San Jose CA	1-800-348-7645
H Plus GCL	Boston MA	1-800-388-9135
Hadley Industries	Luddington MI	1-800-345-4227
Hall Kimbrell Environmental Services In	Lawrence KS	1-800-828-3282
Hall Kimbrell Environmental Services In	Lawrence KS	1-800-346-2860
Hardin Huber Inc.	Greensboro NC	1-800-852-0277
Harding Lawson Associates	Novato CA	1-800-826-0682
Harding Lawson Associates	Novato CA	1-800-578-0821
Harmon Engineering Associates	Auburn AL	1-800-325-0011
Harper & Shuman Inc.	Cambridge MA	1-800-872-4050
Hau ML & Associates	Port Clinton OH	1-800-423-6163
Hawk Engineering	Binghamton NY	1-800945HAWK

Hawthorn Brothers Tree Service Inc.	Bedford Hills NY	1-800-235-7035
Hazardous Environment Life Protection	Little Falls NJ	1-800-554-3577
Hazardous Environment Life Protection	Little Falls NJ	1-800-554-3577
Hazardous Materials Services	Fairfield OH	1-800-846-2925
Hazmat Environmental Group Inc.	Buffalo NY	1-800-836-7800
Hazmat TISI	Columbia MD	1-800-777-TISI
Health & Safety Management	Beaumont TX	1-800-458-4094
Health & Safety Services	Salt Lake City UT	1-800-347-0557
Heart Information Systems Inc.	Schenectady NY	1-800-572-5810
Heath Consultants Inc.	Stoughton MA	1-800-432-8487
Henkle-Buchanan Group	Reno NV	1-800-572-9798
Heraus DSET Laboratories Inc.	Phoenix AZ	1-800-255-3738
Heritage Environmental Services Inc.	Indianapolis IN	1-800-827-0476
Hill International Inc.	Willingboro NJ	1-800-222-0127
Hillman Environmental Co	Cherry Hill NJ	1-800-851-4326
Hillman Environmental Company	Voorhees NJ	1-800-851-4326
Hillman Environmental Company	Union NJ	1-800-232-4326
Hillman Environmental Company	San Mateo CA	1-800-442-4326
Home Environmental Lab	Seaford NY	1-800-626-9912
Horizon Environmental Group Inc.	Cincinnati OH	1-800-326-5721
Horizon Products	Wilmot SD	1-800-872-7573
Howard Smith Screen Co. Inc.	Houston TX	1-800-527-4772
HOWSAFE	Macedonia OH	1-800-966-1133
Hughes Beard Co Inc	Houston TX	1-800-833-3281
Hunter Environmental Science & Engineer	Plymouth Meeting PA	1-800-437-7272
Hunter Environmental Science & Engineer	Raleigh NC	1-800-638-4868
Hydro Group Inc.	Bridgewater NJ	1-800-331-0249
Hydro Search	Reno NV	1-800-347-4937
Hydro-Seach Inc	Golden CO	1-800-544-5528
Hydro-Search Inc.	Irvine CA	1-800-882-8123
Hydrologic	Eudora KS	1-800-495-6442
Hydrologic Inc	Asheville NC	1-800-344-5759
HzW Environmental Consultants	Mentor OH	1-800CONSULT
I E M Sealand Inc	Arlington VA	1-800-448-0876
I E S M C Inc	Crystal Lake IL	1-800-455-1762
I L F C Inc	Rio Rancho NM	1-800-237-4532
I S S Indoor Environmental Services	Lyndhurst NJ	1-800-247-3973
Idaho Nets	Teton ID	1-800-334-6387
IEM Sealand Inc.	Arlington VA	1-800-448-0876
ILFC	Rio Ranch NM	1-800-237-4532
IML		1-800-828-1407
Indoor Air Sciences Associates	Carlsbad CA	1-800-645-8647
Indoor Environmental Solutions	Winston Salem NC	1-800-437-7873
Industrial Hygiene & Safety Technology	Carrollton TX	1-800-562-3617
Inorganic Ventures	Lakewood NJ	1-800-669-6799
Institute For Cooperation Environmental	Philadelphia PA	1-800-875-9470
Institute for Environmental Assessment	Mankato MN	1-800-872-1260
Institute for Environmental Assessments	Anoka MN	1-800-233-9513
Intergraph Corp	Huntsville AL	1-800-826-3515
Intermountain Technical Services Inc.	Grand Junction CO	1-800-477-1835
International Dismantling Machinery (IDM)	South River NJ	1-800-872-1171
International Technology Corp.	Torrance CA	1-800-421-5574
Interphase Environmental	San Diego CA	1-800-457-3300
Invironment	Itasca IL	1-800-722-9093
Invisible Gardner - Organic Consultant	Malibu CA	1-800-354-9296
INX Corp (Indoor Env Qual)	Bernville PA	1-800-786-6254
IPT Corp.	Palo Alto CA	1-800-944-5468
ISS Indoor Environmental Services	Lyndhurst NJ	1-800-247-3973

IT Corp.	Wilmington CA	1-800-421-5574
IT Corp.		1-800-262-1900
J & M Laboratories	Louisville KY	1-800-473-6173
J H Water Systems Inc	Irwin PA	1-800-533-6564
J. J. Keller & Associates Inc.	Neenah WI	1-800-327-6868
J. K Petroleum Equipment	Kansas City KS	1-800-242-4676
J. T Baker Inc.	Phillipsburg NJ	1-800-582-2537
JABA Inc.	Tucson AZ	1-800-999-2348
JACA Corp.	Fort Washington PA	1-800-292-2510
Jamestown Marine Services	Jamestown RI	1-800-332-0100
Jason Associates (Wood Products)	Fort Collins CO	1-800-777-7653
JMM Operational Services	Denver CO	1-800-234-4566
Joiner Tom & Associates	Tuscaloosa AL	1-800-226-2311
Joint Laboratory Services	Ann Arbor MI	1-800-227-5574
Jones Environmental Technology Inc.	Austin TX	1-800-645-6637
Jordan Systems Inc.	Cedar Rapids IA	1-800-859-3023
Joy Environmental Technology Inc	Houston TX	1-800-528-1159
Joyce Environmental Consultants	Orlando FL	1-800-533-1565
Joyce Environmental Consultants	Hastings FL	1-800-447-6820
Karden Associates Inc	Seattle WA	1-800-892-7908
KBM Forestry Consultants Inc.	Thunder Bay ON CN	1-800-465-3001
KBN Engineering & Applied Sciences Inc.	Gainesville FL	1-800-333-4526
KDF Fluid Treatment Inc.	Constantine MI	1-800-437-2745
KEI Industrial Services	Morrisville PA	1-800-447-3534
Kelleher Environmental Inc	Burnsville MN	1-800-553-2648
Kemron Environmental Services Inc.	McLean VA	1-800-777-1042
Kern Environmental Services	Bakersfield CA	1-800-332-5376
Kern Environmental Services	Bakersfield CA	1-800-332-5376
Keystone Environmental Resources Inc.	Monroeville PA	1-800-648-9001
KHI Limited	Oceanside CA	1-800-223-3642
Killam Associates	Meeker NJ	1-800-832-3272
King D G Associates Planners	Rancho Cucamonga CA	1-800-729-9591
Kleenco Limited	Marshfield WI	1-800-952-7507
KLS & Associates	Altamonte Springs FL	1-800-682-9783
Knoblock's Environmental	Adamstown PA	1-800-446-5288
KW Brown Environmental Services	Houston TX	1-800-749-9280
KWS Group Inc.	Kenosha WI	1-800-532-7394
L A W E Environmental Associates	Newburyport MA	1-800-255-9265
L S & A Environmental	New York NY	1-800-331-0312
La Velle Enterprises	Oceanside CA	1-800-722-0120
Laidlaw Environmental Services	Laurel MD	1-800-638-4440
Laidlaw Environmental Services Inc.	Columbia SC	1-800-356-8570
Laidlaw Environmental Services Inc.	Baton Rouge LA	1-800-232-5374
Lakeshore Environmental	Grand Haven MI	1-800-844-5050
Land Management Services	Northeast Heights MD	1-800-336-5850
Landa Inc.	Portland OR	1-800-547-8672
Landauer Inc.	Glenwood IL	1-800-528-8327
Larks Engineering/Consulting (OSHA)	Houston TX	1-800-594-1345
Layne Inc.	Mission Woods KS	1-800-433-6776
LCH Construction Management	Snellville GA	1-800-982-6467
Leader Industries	Portage IN	1-800-437-6122
Leberco Testing Inc.	Roselle Park NJ	1-800-523-5227
Ledford Engineering Inc.	Ashville NC	1-800-654-8891
Lincoln Enviromental	Smithfield RI	1-800-659-3353
Lockwood Greene Engineers Inc.	Spartanburg SC	1-800-845-3302
Longyear US Products Group	Doraville GA	1-800-922-9497
Lycott Environmental Research Inc.	Southbridge MA	1-800-462-8211
M L Hau & Associates Inc	Port Clinton OH	1-800-423-6163

Mabbett, Capaccio & Associates Inc.	Bedford MA	1-800-877-6050
MAC Environmental	Sabetha KS	1-800-223-2191
Mackinac Environmental Inc.	St Ignace MI	1-800-582-8806
Maecorp. Inc.	Grand Rapids MI	1-800-382-2769
Maloney Site	Middlefield CT	1-800-453-6021
Manteq Engineers	Lake Jackson TX	1-800-633-4473
Marc Power Washers	Mira Loma CA	1-800-338-2183
Marietta Environmental Services	Waterford OH	1-800-582-2531
Maumee Research & Engineering Inc.	Perrysburg OH	1-800-874-3882
McComas Technologies Inc.	Erlanger KY	1-800-437-6785
McGinley Associates P.A.	Stillwater MN	1-800-879-9231
McLaren/ Hart Environmental	Lester PA	1-800-645-6741
Medical Gas Control	Apex NC	1-800-822-2595
MEMA Environmental Institute/Filter Hot	Research Triangle Park N	1-800-993-4583
METCO Environmental	Addison TX	1-800-394-1194
Meteorological and Environmental Planni	Markham ON CN	1-800-387-9729
Meteorological Evaluation Services	Amityville NY	1-800-952-2052
MGM Petro Equipment & Env Ser	Lakeland FL	1-800-624-7224
Michigan Pumping Service	Trenton MI	1-800-551-5774
Micro-Fiber Laboratories Inc.	Northbrook IL	1-800-373-LABS
Microbial Environmental Services Inc.	Des Moines IA	1-800-328-3096
Mid-South Forestry Inc.	New Albany MS	1-800-238-9884
Mid-State Associates Environmental Serv	Baraboo WI	1-800-228-3012
Midpacific Environmental Lab	Mountain View CA	1-800-292-6735
Midstates Environmental Remediation Ser	Ft Wayne IN	1-800-526-3774
Minnesota Technical Assistance Program	Minneapolis MN	1-800-247-0015
Minnesota Technical Assistance Program	Minneapolis MN	1-800-247-0015
Montgomery Watson	Pasadena CA	1-800-427-4457
Mostardi Platt Associates Inc	Elmhurst IL	1-800-445-2367
Mostardi Platt Associates Inc.	Bensenville IL	1-800-445-2367
MSE Inc.	Butte MT	1-800-441-8213
Munchiando Excavating	Wheat Ridge CO	1-800-662-0979
Munchiando Excavating	Denver CO	1-800-662-0979
Nat'L Medical Advisory Svc	Bethesda MD	1-800-258-0014
NATEC Of Texas	Houston TX	1-800-446-2832
Natec of Texas	Houston TX	1-800-446-2832
National Energy Consultants	Cedar Rapids IA	1-800-373-8805
National Environmental Service Co.	Tulsa OK	1-800-328-8335
National Hospital Management Systems	La Grange GA	1-800-292-9323
National Hospital Management Systems	La Grange GA	1-800-292-9323
National Laboratories Inc.	Evansville IN	1-800-444-4119
National Medical Advisory Service	Bethesda MD	1-800-258-0014
National Recycling Technology	Boca Raton FL	1-800-541-0400
National Society Of Consulting Soil Sci	Elma NY	1-800-535-7148
National Society of Consulting Soil Sci	Washington DC	1-800-535-7148
National Sorbents Inc.	Cincinnati OH	1-800-677-9465
Natural Resources Research Institute	Duluth MN	1-800-562-0004
Nedco Limited	Lyndhurst NJ	1-800-858-6226
Nedco Limited	Lyndhurst NJ	1-800-858-6226
Neim The Green Store	Burlington VT	1-800-551-8086
Ness Doug for Esstek	Meridian ID	1-800-346-5161
Network Environmental Systems Inc. (NES)	Folsom CA	1-800-658-8493
New York Testing Laboratories Inc.	Westbury NY	1-800-433-0008
North American Environmental Service	Buffalo NY	1-800-876-2903
North American Green Inc	Evansville IN	1-800-448-2040
North American Pollution Control System	Bethlehem PA	1-800-752-0237
North American Pollution Control Systems		1-800-231-4677
North American Pollution Control System	Linden NJ	1-800-752-0237

North American Software Inc.	Tustin CA	1-800-966-5678
North American Weather Consultants	Salt Lake City UT	1-800-658-8493
Northeastern Analytical Corp	Marlton NJ	1-800-622-5080
Northeastern Analytical Corp.	Marlton NJ	1-800-622-5080
Northern Environmental Decontamination	McComb OH	1-800-334-8101
Northwest Texas Environment Training	Amarillo TX	1-800-594-0008
NU-SOILS	Parsippany NJ	1-800-225-7645
Nuchemco	Annandale VA	1-800-682-4362
Nuchemco	Annandale VA	1-800-682-4362
NUCON International Inc.	Columbus OH	1-800-992-5192
NUS Corp	Houston TX	1-800-262-3027
NUS Training Corp.	Gaithersburg MD	1-800-338-1505
Nutech Enterprises	Oceanside CA	1-800-722-8896
Nutech Enterprises	Oceanside CA	1-800-722-8896
OBG Operations	Syracuse NY	1-800-435-0797
Occupational & Environmental Diseases -	Lansing MI	1-800-446-7805
Occusafe Inc.	Wheeling IL	1-800-323-7597
Ogden Environmental and Energy Services	Huntsville AL	1-800-541-9447
OH Materials	Trenton NJ	1-800-327-2853
OH Materials	Princeton NJ	1-800-327-2853
OHM	Findlay OH	1-800-537-9540
OHM	N Tonawanda NY	1-800-457-4412
OHM	Clarence Center NY	1-800-457-4412
OHM Corp.	Findlay OH	1-800-537-9540
OHM Environmental	Lanham MD	1-800-662-7618
OHM Environmental	Norristown PA	1-800-448-5991
OMC Environmental	Lanham MD	1-800-662-7618
OMC Environmental	Norristown PA	1-800-448-5991
Omega Engineering Inc.	Stamford CT	1-800-826-6342
OMNI Environmental Services	Durham NC	1-800-843-6066
Op-Tech Environmental Services Inc.	Massena NY	1-800-225-6750
Optronic Laboratories Inc.	Orlando FL	1-800-899-3171
Orbis	Exton PA	1-800-243-5559
Orin Environmental Inc	Hudson OH	1-800-272-2819
OSHA Training Maple Woods Community	ColKansas City MO	1-800-841-7158
Otto H. York Co. Inc.	Parsippany NJ	1-800-524-1543
P E R S Inc.	Ponca City OK	1-800-328-2482
P T & L Environmental Services	Paramus NJ	1-800-486-4509
Pacific Environmental Services Co.	Port Townshend WA	1-800-222-9219
Pacific Tree Care	Calistoga CA	1-800-743-0361
Packard Consultants Inc.	Bellingham MA	1-800-732-7357
Parkham Industrial Distributors	Louisville KY	1-800-255-6073
Parkland Lab	Grandview IL	1-800-541-9568
Parsons Environmental Services Inc.	Pasadena CA	1-800-883-7300
Parsons, Bromfield - Redniss & Mead Inc	Stamford CT	1-800-327-2060
Pelican Environmental Products	Katonah NY	1-800-345-9906
Peninsu-Lab (Pest Management)	Poulsbo, WA	1-800-635-6866
Pennsylvania Environmental Resources -	Gilbertsville PA	1-800-237-2366
Permit Consultants	Brownwood TX	1-800-351-1363
Petro Environmental Services	Harrisburg PA	1-800-438-7374
Petro Global Consultants Of San Antonio	San Antonio TX	1-800-528-6434
Phase One Inc	Aliso Viejo CA	1-800-774-2731
Phase One Inc.	Aliso Viejo CA	1-800-524-8877
Piedmont Environmental	Kannopolis NC	1-800-235-3006
Piedmont Environmental Service	Greensboro NC	1-800-572-7295
Pinnacle Environmental Group Inc	Toledo OH	1-800-828-9096
Planning Resources Inc.	Wheaton IL	1-800-255-4354
Pollution Control Inc.	Florence KY	1-800-446-4724

Pollution Enterprises	Bridgewater NJ	1-800-854-0410
Pollution Enterprises Inc.	Horsham PA	1-800-253-7248
Powers Elevation Co. Inc.	Aurora CO	1-800-824-2550
Prairie Environmental	Plainfield IL	1-800-536-4088
Precision Air Inc.	Wilmington DE	1-800-851-9703
Precision Environmental Inc.	Raleigh NC	1-800-334-8016
PRENCO	Matthews NC	1-800-245-8579
Princeton Testing Laboratory	Princeton NJ	1-800-548-TEST
Princeton Testing Laboratory Inc	Princeton NJ	1-800-548-8378
Process Data Control	Dallas TX	1-800-472-7848
Production Management for Crops	Fremont NW	1-800-736-7870
ProEco Inc.	Tampa FL	1-800-788-5103
Professional Analytical & Consulting Services	1-800-367-2587	
Professional Analytical And Consulting	Coraopolis PA	1-800-367-2587
Professional Service Industries	Lawrence KS	1-800-548-7901
Professional Service Industries	Lawrence KS	1-800-548-7901
Professional Tank Services Ltd	Des Plaines IL	1-800-624-9223
Professional Tank Services Ltd.	Schiller Park IL	1-800-624-9223
Progressive Environmental & Safety	Bonner Springs KS	1-800-528-8587
Progressive Environmental And Safety	Bonner Springs KS	1-800-528-8587
Project Support Inc.	Portland OR	1-800-453-1562
Proline Enterprises	Albert Lea MN	1-800-845-4346
ProPark Group Inc.	Margate FL	1-800-388-7212
Prudential Rod Girtman Inc.	Portland OR	1-800-766-7346
PSC Engineers & Consultants Inc.	Limerick PA	1-800-969-2579
PSC Environmental	Limerick PA	1-800-824-6126
PTS Environmental Engineering	Loomis CA	1-800-962-1069
Pullman Industrial Systems	Laguna Hills CA	1-800-262-6914
Pure Choice Inc.	Edina MN	1-800-845-5544
PW Environmental	Santa Paula CA	1-800-334-8911
Q Source	Miamisburg OH	1-800-356-9039
QED Environmental Systems	Ann Arbor MI	1-800-624-2026
Qsource	Miamisburg OH	1-800-356-9039
Quadrex Environmental Co.	Gainesville FL	1-800-947-2829
Quaker Chemical Corp	Conroe TX	1-800-327-8934
Quaker Chemical Corp	Ojai CA	1-800-258-9411
Quaker Chemical Corp.	Ojai CA	1-800-258-9411
Quaker Chemical Corp.	Conroe TX	1-800-327-8934
Qualimetrics Inc.	Sacramento CA	1-800-824-5873
Quality Air Division	Romeoville IL	1-800-247-7769
Quest Environmental Resources Corp	Indianapolis IN	1-800-878-4872
R E R C Environmental	Dallas TX	1-800-572-5485
R-C Environmental Services & Technologi	Somerville NJ	1-800-722-3780
Rader Environmental Services	Findlay OH	1-800-858-7374
Rader Environmental Services	Findlay OH	1-800-858-7374
Ranson Environmental Consultants Inc.	Newburyport MA	1-800-439-1822
Reach Associates	S Orange NJ	1-800-246-9628
REACT Environmental Engineers	St Louis MO	1-800-325-1398
Reactives Management Corp	Chesapeake VA	1-800-372-6742
Recovery Equip Supply Inc	Maple Grove MN	1-800-541-0518
Recovery Equipment Supply Inc.	Maple Grove MN	1-800-541-0518
Recra Environmental	Amherst NY	1-800-52RECRA
Red Rose Environmental	Wyomissing PA	1-800-437-4696
Refuse & Environmental Systems Inc.	W Springfield MA	1-800-443-1924
Relco Engineers	Sante Fe Springs CA	1-800-959-0894
Reliable Tank Technicians	Bangor PA	1-800-922-2318
Resource Chemical Company	Schererville NC	1-800-457-3282
Resource Recovery of America	Mulberry FL	1-800-752-3242

Restaurant Environmental Services	Ft Myers FL	1-800-238-0487
Rhone-Poulenc Basic Chemicals Co.	Shelton CT	1-800-521-2289
Riedel Environmental Services Inc.	Portland OR	1-800-334-0004
RMC Medical	Philadelphia PA	1-800-959-0284
Robert D. Niehaus Inc. (RDN)	Goleta CA	1-800-350-4888
Rogers Entomological Services Inc.	Cleveland MS	1-800-748-8727
Rollins Chempak Inc	Livonia MI	1-800-435-4838
Rollins Chempak Inc.	Livonia MI	1-800-435-4838
Rollins Environmental Services	Wilmington DE	1-800-992-7837
Roth Asbestos and Environmental Consult	Westwood KS	1-800-279-7220
Roux Associates Inc.	Huntington NY	1-800-322ROUX
Royce Instrument Corp.	New Orleans LA	1-800-347-3505
Rust Environment & Infrastructure Inc.	Greenville SC	1-800-868-0373
Rust Environmental	Sheboygan WI	1-800-868-0373
Rust International	Birmingham AL	1-800-247-3122
S & Me	Raleigh NC	1-800-662-7454
S B L Group	Marietta GA	1-800-892-2090
S Q G Management Corp	Slidell LA	1-800-554-3317
Safe Air	N Canton OH	1-800-245-9247
Safety & Compliance Associates Inc.	Roanoke VA	1-800-283-0084
Safety & Environmental Technologies	San Diego CA	1-800-821-9191
Safety Specialists	Santa Clara CA	1-800-421-6710
Safewaste Inc.	Wheeling IL	1-800-323-7597
Sagian Inc.	Indianapolis IN	1-800-352-4975
Sahara Group The	Anaheim CA	1-800-752-1818
Sanborn	Wrentham MA	1-800-343-3381
Sand Sifters Beach Cleaning	Sayville NY	1-800-642-0083
Sand Sifters Beach Cleaning	Sayville NY	1-800-642-0084
Sargent & Lundy	Chicago IL	1-800-438-7208
Sassi Training Systems Inc	Auburn AL	1-800-633-5471
Sassi Training Systems Inc.	Auburn IL	1-800-633-5471
Schaible Associates	Mt Joy PA	1-800-832-5564
Schaible Associates	Mt Joy PA	1-800-832-5564
Science Applications Intl	Golden CO	1-800-270-7242
Scientific Software-Intercomp Inc.	Denver CO	1-800-525-5819
Scott Allard & Bohannon	Phoenix AZ	1-800-253-0004
Select Environmental Technologies Inc	Marysville PA	1-800-547-1068
Select Environmental Technologies Inc.	Marysville PA	1-800-547-1068
Selective Settlements International	Portland OR	1-800-288-7005
Seneca Environmental Services	Davenport IA	1-800-356-2522
Sergent, Hauskins & Beckwith Engineers	Phoenix AZ	1-800-248-2472
Shirley Environmental Testing	Grand Island NE	1-800-292-3490
SHRED-TECH Ltd	Cambridge ON CN	1-800-465-3214
Sigma Environmental Services	Oak Creek WI	1-800-365-3840
Simon Hydro Search	Houston TX	1-800-548-7667
Simon Hydro-Search	Golden CO	1-800-544-5528
Simon Hydro-Search	Houston TX	1-800-548-7667
Simon Hydro-Search	Irvine CA	1-800-882-8123
Simon Hydro-Search Inc	Huntington Beach CA	1-800-882-8123
Site Environmental Services Inc	Orland Park IL	1-800-750-7483
Site Environmental Services Inc.	Orland Park IL	1-800-750-7483
Site Scan	Marietta OH	1-800-752-5366
Skelly & Loy Inc.	Harrisburg PA	1-800-892-6532
SLC Consultants/Constructors Inc.	Lockport NY	1-800-537-8203
Soil Consultants of Maryland Inc.	Clinton MD	1-800-441-7224
Soil Remediation	Denver CO	1-800-441-1968
Soilwater Investigations	Albuquerque NM	1-800-242-6505
Soilwater Investigations	Albuquereque NM	1-800-242-6505

Solvent Recovery Systems Inc.	Huffman TX	1-800-367-5773
Sommers Consultants	Port Washington NY	1-800-862-3838
Soresi Chemical Group	Morrisville PA	1-800-922-0407
Southdown Environmental Systems	Franklin TN	1-800-367-5456
Southeatern Testing Lab	Leighton AL	1-800-334-1971
Southern Ecology Management	Corpus Christi TX	1-800-221-0206
Southern Services Inc.	Panama City Beach FL	1-800-852-8878
SoyPro International Inc. (oilseed)	Cedar Falls IA	1-800-747-4706
Special Resource Management	Boise ID	1-800-654-2504
Special Resource Management	Butte MT	1-800-735-8964
Special Resource Management	Butte MT	1-800-334-8911
Special Resource Management	Cheyenne WY	1-800-237-2647
Special Resource Management	Fargo ND	1-800-445-4404
Special Resource Management	Pierre SD	1-800-822-0287
Specialty Conveyance Services Inc.	Ft Wayne IN	1-800-526-3774
Spectrum Labs	Davenport IA	1-800-851-5316
Spectrum Labs	Davenport IA	1-800-851-5316
SQG Management Corp.	Slidell LA	1-800-554-3317
SRS Environmental Inc.	Tampa FL	1-800-282-0832
Sta-Clean Products/Environmentally Safe	Bakersfield CA	1-800-825-3464
Stan A. Humber Consultants Inc. (SAHCI)	New Lenox IL	1-800-383-0468
Standard Testing and Engineering Co.	Oklahoma City OK	1-800-876-0452
State Environmental Management	Wrightwood CA	1-800-638-1587
State Environmental Management	Wrightwood CA	1-800-638-1587
Stoller Corp	Carlsbad NM	1-800-748-3131
Stone and Webster Environmental Technol	Boston MA	1-800-421-3042
Stone Environmental Services	Hampstead NH	1-800-639-4503
Storch Radon	Fall River MA	1-800-362-6290
Structure Probe Inc.	W Chester PA	1-800-2424-SPI
STS Consultants Ltd.	Northbrook IL	1-800-859-STS1
Sunshine Makers Inc.	Huntington Harbour CA	1-800-228-0709
Superior Training	Lake Charles LA	1-800-362-9781
Surface Science Laboratories	Mountain View CA	1-800-321-4775
Sverdup Civil Inc.	Maryland Heights MO	1-800-325-7910
Sybron Biochemical	Birmingham NJ	1-800-678-0020
Sybron Chemicals Inc. Bioremediation Se	Palm Harbor FL	1-800-645-6809
Synergic Resources Corp	Bala Cynwood PA	1-800-544-2522
SZ Mansdorf & Associates	Cuyahoga Falls OH	1-800-331-3044
Taplin Enterprises	Kalamazoo MI	1-800-325-6720
Taplin Enterprises	Kalamazoo MI	1-800-325-6720
Taralan Corp (Agri)	Geneva IL	1-800-776-3823
Target Environmental	Vineland NJ	1-800-428-6017
Team Environmental Services	Bangor ME	1-800-662-8326
Team Inc.	Alvin TX	1-800-662-8326
Tenco Environmental Laboratories	Schererville IN	1-800-428-3311
Terra Technologies	Houston TX	1-800-458-3772
Terracon Environmental Inc.	Lenexa TX	1-800-593-7777
Terraine Environmental Services	Knoxville TN	1-800-531-1242
TerraTech Environmental Services Inc.	Wichita KS	1-800-776-9565
The Altus Group	Arlington TX	1-800-247-5438
The Clearwater Group	Bend OR	1-800-285-3668
The Consortium Env and Occupa	Cincinnati OH	1-800-626-4174
The Emissions Measurement People	Canon City CO	1-800-222-4187
The Enviramigo	Madisonville LA	1-800-554-2012
The Environmental Appraisers	Corpus Christi TX	1-800-551-2532
The Envirovision Group Inc.	Congers NY	1-800-989-7483
The ERM Group	Exton PA	1-800-544-3117
The Healing Earth Company	Nevada City CA	1-800-257-5723

The Safety Specialist Inc.	Naperville IL	1-800-635-7471
The Wings Co.	Rosemead CA	1-800-821-8732
Think Tank Resources Inc.	Fairfax VA	1-800USACLEM
TN Technologies Inc.	Round Rock TX	1-800-736-0801
Total Recovery	Parker Ford PA	1-800-258-5585
Tower Environmental Inc	Tampa FL	1-800-522-2241
Toxico	Southfield MI	1-800-227-7906
TPS Technologies Inc	Apopka FL	1-800-940-2666
Tracer Research Corp.	Tucson AZ	1-800-989-9929
Trap Zap Environmental Systems	S Norwalk CT	1-800-282-8727
TRC Environmental Consultants	E Hartford CT	1-800-367-1044
TRC Environmental Corp.	Windsdor CT	1-800-TRC-5601
TRC Environmental Corp.		1-800-365-8254
Treatek Inc.	Grand Island NE	1-800-833-3335
Triangle Environmental Services	Research Triangle Park NC	1-800-367-4862
Tricon Environmental Inc	Auburn AL	1-800-854-7917
Tricon Environmental Inc.	Auburn AL	1-800-854-4312
Tricon Environmental Inc.	Birmingham AL	1-800-854-7917
Trident Environmental Services	Summerville SC	1-800-344-8371
Trident Environmental Services Inc	Summerville SC	1-800-344-8371
TTI Environmental Inc.	Marlton NJ	1-800-228-8003
Turnkey Environmental Consultants	Mt Prospect IL	1-800-451-7423
U S Ecology	Louisville KY	1-800-626-5334
U S Testing Co.	Hoboken NJ	1-800-777-8378
Unisource	Houston TX	1-800-247-2004
United Bio-Tek	Walterboro SC	1-800-982-7324
United States Environmental Consulting	Anderson IN	1-800-245-1247
United States Environmental Consulting	Troy MI	1-800-245-0247
URS Consultants Inc.	New York NY	1-800-327-8877
US Environmental Consulting Inc	Anderson IN	1-800-245-1247
US Tank Management Inc.	Albuquerque NM	1-800-786-8796
USPCI	Houston TX	1-800-877-2401
USPCI Remedial Services	Boulder CO	1-800-877-2416
Ustnet	Santee CA	1-800-382-2778
Vaper Extraction Technology	San Clemente CA	1-800-524-9172
Versar Inc.	Springfield VA	1-800-283-7727
Vista Environmental Information	San Diego CA	1-800-733-7606
Volz Environmental Services	Pittsburgh PA	1-800-433-9709
W L B J	Raymore MO	1-800-874-4501
W. F. Biggins Associates Inc.	E. Longmeadow MA	1-800-722-3563
Waetzman Planning Group	Ardmore PA	1-800-669-5202
Walker Carl Engineers Inc.	Kalamazoo MI	1-800-FYIPARK
Wapora Inc.	McLean VA	1-800-777-1042
Ward's Environmental Services Co	Bronx NY	1-800-438-9157
Warzyn Inc.	Madison WI	1-800-388-0288
Waste Control Services	Channelview TX	1-800-832-7536
Waste Energy Technology Inc.	Ft Walton Beach FL	1-800-441-6822
Waste Management Inc. ES & H Hotline	Oak Brook IL	1-800-323-2435
Waste Research & Recovery Inc.	Atlanta GA	1-800-336-1591
Waste Tron	Charleston WV	1-800-352-1146
Waste Tron	Wheelersburg OH	1-800-352-6144
Water & Air Research	Gainesville FL	1-800-242-4927
Water & Oil Technologies Inc.	Montgomery IL	1-800-841-6580
Water & Waste Environmental Services	Mishawaka IN	1-800-452-8556
Waveguard International	Austin TX	1-800-552-9283
Wedding & Associates	Ft Collins CO	1-800-367-7610
Wenck Associates Inc.	Maple Plain MN	1-800-472-2232
West Central Environmental Consultants	Morris MN	1-800-422-8356

Western Environmental Services and Test	Casper WY	1-800-545-5711
Westester	Incline Village NV	1-800-222-9168
Westinghouse Environmental	Decatur GA	1-800-752-3303
Westinghouse Groundwater Recovery	Doraville GA	1-800-922-9497
Westinghouse Haztech Inc Emergency Resp	Atlanta GA	1-800-358-4135
Westinghouse Haztech Inc. Emergency Res	Decatur GA	1-800-358-4135
Westinghouse Remediation Services	Clarkston GA	1-800-543-3213
Weston and Sampson Engineers Inc.	Peabody MA	1-800-726-7766
Weston Geophysical	Westborough MA	1-800-334-8011
Williams Environmental Services Inc.	Stone Mountain GA	1-800-247-4030
Winton Lab	Plymouth Meeting PA	1-800-843-5307
Wolfe & Associates	Winston Salem NC	1-800-982-0088
Woodward Clyde Consultants	Denver CO	1-800-776-3296
World 2000 Environmental Services Inc.	Lighthouse Point FL	1-800-932-9652
World Advanced Technology	Mt Clemens MI	1-800-321-9280
Wright Environmental Service	Columbus GA	1-800-562-2847
Xcel Environmental	Providence RI	1-800-638-9235
Yank A Tank	Anaheim CA	1-800-282-9265
Yank A Tank	Yorba Linda CA	1-800-282-9265
Zimpro Passavant	Avon Lake OH	1-800-421-3144
Zimpro Passavant	Irondale AL	1-800-633-9501
Zimpro Passavant	Rothschild WI	1-800-826-1476

ENVIRONMENTAL INSURANCE

National Assurance Corp	Greenville SC	1-800-736-1902
Rose-Tillman	Edwardsville IL	1-800-228-3328
Terra Insurance Company	Corte Madeira CA	1-800-872-0077

ENVIRONMENTAL PRODUCTS

Alpine Environmental Inc	Mt Vernon WA	1-800-655-5990
Anderson Affiliates-Enzymes Plus Divisi	Sugar Land TX	1-800-444-7741
Anoroc Scientific	Moonachie NJ	1-800-777-3817
Bio Alliance	Brea CA	1-800-246-2677
BioEnviroTech Inc	Tomball TX	1-800-758-3254
Biosource Inc	Silsbee TX	1-800-643-5250
Blue Rhubarb-Ecosac	Cambria CA	1-800-926-1017
Canaan Environmental Products Inc	Lilburn GA	1-800-842-1088
Dexsil Corp	Hamden CT	1-800-433-9745
Diatec Environmental	Batavia IL	1-800-854-7952
Ecolo Kids	Topeka KS	1-800-423-7202
Environmental Chemical	Rogers TX	1-800-818-8102
Environmental Distributors	Soddy Daisy TN	1-800-219-2121
Environmental Products & Supplies	Berwyn Heights MD	1-800-260-4320
Four States Tank	Evansville IN	1-800-467-3075
Fuel Dynamics Inc	Holly Hill SC	1-800-838-6766
Future Best	Southampton PA	1-800-626-5889
Guatex Sales	San Antonio TX	1-800-760-4998
Guthrie Associates Inc	Annandale NJ	1-800-758-4361
Horizon Products Inc	Wilmot SD	1-800-872-7573
Keck Instruments	Williamston MI	1-800-542-5681
Keystone Environmental Services & Produ	Stroudsburg PA	1-800-997-7455
M & B Enterprise	Dallas TX	1-800-761-0074
Natures Technology	Eastland TX	1-800-213-7063
Organic Technologies Inc	Miami FL	1-800-349-9900
Paragon Environmental Systems	Escondido CA	1-800-985-0055
Pyramid Drilling Supply & Mfg	Londonderry NH	1-800-434-4254

R S T Cable & Tape	Ronkonkoma NY	1-800-875-8273
Resource Planning Associates	Cornell NY	1-800-432-7721
Roach Master	Woodland Hills CA	1-800-847-6224
Shaklee Authorized Distributor	Collinsville OK	1-800-574-3155
Southern Stud Weld/Inspection Port	Irving TX	1-800-929-0732
Southern Stud Weld	Houston TX	1-800-929-0296
Tri Synergy	Escondido CA	1-800-446-6076
United Environmental	Hampton VA	1-800-597-2847
Universal Sensors & Devices Inc	Chatsworth CA	1-800-899-7121
Verde Environmental	Houston TX	1-800-626-6598

ENVIRONMENTAL RECRUITERS & EMPLOYMENT INFORMATION

Bailey Associates	Monroe CT	1-800-627-2712
Career Connection	Thousand Oaks CA	1-800-266-3010
F W Dodge	New York NY	1-800-541-9913
The Green Book Report	Andover MA	1-800-527-3304
Phillips International Inc.	Greenville SC	1-800-638-1661

FISH HATCHERIES

Aylesworth Fish & Bait	St Petersburg FL	1-800-227-4577
Florida Fresh Fisheries	Punta Gorda FL	1-800-476-3474
Lonoke Fish Farm	Lonoke AR	1-800-362-5035

FORESTRY CONSULTANTS

Carter John A Forest Consultants	Bedford VA	1-800-451-1141
Crawford John Consulting Forester	Nacogdoches TX	1-800-874-4259
Dude Herring Reforestation Svcs	Ruston LA	1-800-253-6796
Forest Partnership The	Burlington VT	1-800-858-6230
Independent Timber Consultants-ITC	Sedro Wooley WA	1-800-952-7540
John A Carter Forest Consultants	Bedford VA	1-800-451-1141
Mark King Forestry Consultant Service	Winona MS	1-800-858-5464
Maurice Williamson/ACF Forestry Consult	Colville WA	1-800-435-9339
Penn Forestry Co. Inc.	Biglerville PA	1-800-327-4772
Rhodes Forestry Services Inc.	Aiken SC	1-800-348-5263
Shawn Montee Forrestry Services	Coeur d'Alene ID	1-800-664-9821
Southern Forestry Consultants	Bainbridge GA	1-800-262-8182
The Forest Partnership	Burlington VT	1-800-858-6230
WW Sellers & Associates	Ramer AL	1-800-451-1354

GEOLOGY, GROUND WATER & HYDROGEOLOGY

Aaron Environmental	Waterbury CT	1-800-394-2374
ACS Industries Inc.	Houston TX	1-800-231-0077
Advanced Geologic Exploration	Reno NV	1-800-852-2960
Aeromix Systems Inc.	Minneapolis MN	1-800-879-3677
American Sigma	Medina NY	1-800-635-4567
Associated Design & Manufacturing	Alexandria VA	1-800-837-8257
Baski Inc.	Denver CO	1-800-522-2754
Brinecell Inc.	Salt Lake City UT	1-800-468-8241
Calgon Carbon Corp.	Pittsburgh PA	1-800-422-7266
Capital Environmental Drilling Services	Dunbarton NH	1-800-924-1192
Carbtrol Corp.	Westport CT	1-800-242-1150
CGS Contract Geological Services	Sparks NV	1-800-247-6853

Chapman RE	W Boylston MA	1-800-727-6231
Clean Environment Equipment	Oakland CA	1-800-537-1767
Clements Associates Inc.	Newton IA	1-800-247-6630
Coleman Energy & Environmental Systems	Golden CO	1-800395HAWK
Condor Geotechnical Services	Wheat Ridge CO	1-800-635-9582
Cook Screen Technologies	Cincinnati OH	1-800-743-9355
Delta Environmental Consultants Inc.	St. Paul MN	1-800-888-1331
E Ryder Well Drilling	Hanover MA	1-800-698-3394
East Coast Environmental Service	Wallingford CT	1-800-451-2835
Enviro Products Inc.	Lansing MI	1-800-368-4764
Enviroscan	Rothschild WI	1-800-338-7226
Envirotech Drilling Inc.	W Bridgewater MA	1-800-583-2680
Forestry Suppliers Inc.	Jackson MS	1-800-647-5368
Gelman Sciences	Ann Arbor MI	1-800-825-4435
Geo-Hydro-Data	Tehacahapi CA	1-800-351-0507
GeoGuard Inc.	Medina NY	1-800-645-7654
Geohazards	Gainesville FL	1-800-770-9990
GeoLogic Recovery Systems	Mulberry FL	1-800-752-3242
Geophysical Survey Systems Inc.	N Salem NH	1-800-893-1109
GeoPure Continental Systems & Services	Gainesville FL	1-800-342-1103
GeoResearch	Long Beach CA	1-800-523-4786
Geotech Environmental Equipment Inc.	Denver CO	1-800-833-7958
Geraughty & Miller Inc.	Denver CO	1-800-225-8419
Groundwater Technology Inc.	Norwood MA	1-800-635-0053
Handar Inc.	Sunnyvale CA	1-800-955-7367
Hargis Associates Inc.	Burbank CA	1-800-554-2744
Howard Smith Screen Co	Houston TX	1-800-527-4772
Hydra-Stop Inc.	Blue Island IL	1-800-538-7867
Hydro Chem Service	San Francisco CA	1-800-822-0019
Hydro Engineering	Salt Lake City UT	1-800-247-8424
Hydro Geo Chem Inc.	Tucson AZ	1-800-727-5547
Hydro Group Inc.	Bridgewater NJ	1-800-331-0249
Hydro Tek	Mokena IL	1-800-526-4955
Hydro-Search Inc.	Huntington Beach CA	1-800-882-8123
Hydro-Services	Missouri City TX	1-800-231-6913
Hydro-Tech Inc.	Prince George VA	1-800-292-3417
Hydrolab Corp	Austin TX	1-800-949-3766
Hydrolab Corp.	Austin TX	1-800-949-3766
HydroLogic Inc.	Frankfort KY	1-800-728-2251
Hydrology Laboratories		1-800-645-1162
In-Situ Inc.	Laramie WY	1-800-446-7488
Instrumentation Northwest Inc.	Redmond WA	1-800-776-9355
Isco Inc., Environmental Div.	Lincoln NE	1-800-228-4373
Keck Instruments	Williamston MI	1-800-542-5681
Kinetico Engineered Systems Inc.	Newbury OH	1-800-633-5530
Knox Technologies	Fredericktown OH	1-800-253-5669
LaFramboise Well Drilling	Thompson CT	1-800-624-2327
Longyear US Products Group	Stone Mountain GA	1-800-241-9468
Marsh-McBirney Inc.	Frederick MD	1-800-368-2723
MDA Scientific Inc.	Lincolnshire Il	1-800-344-4632
Memtek Corp.	Billerica MA	1-800-527-0433
Miller Drilling Co. Inc.	Lawrenceburg TN	1-800-749-5334
Millipore Corp.	Bedford MA	1-800-645-5476
MPC Environmental	Detroit MI	1-800-521-8232
National Ground Water Association	Columbus OH	1-800-551-7379
Nepcco	Ocala FL	1-800-277-3279
ORS Environmental Equipment	Greenville NH	1-800-228-2310
Ozonology Inc.	Evanston IL	1-800-523-5883

PBSJ Inc.	Miami FL	1-800-597-7275
Plains Environmental Services	Salina KS	1-800-542-0445
Product Level Control	Eagan MN	1-800-229-9355
QED Environmental Systems Inc.	Ann Arbor MI	1-800-624-2026
RREM Inc.	Duluth MN	1-800-777-7380
Simon Hydro-Search	Golden CO	1-800-544-5528
Solinst Canada Ltd.	Glen Williams ON CN	1-800-661-2023
Subsurface Investigations Inc.	Congers NY	1-800-984-7483
Sybron Chemicals Inc.	Birmingham NJ	1-800-678-0020
Target Soil Gas Surveys	Dallas TX	1-800-776-3340
Terraplus	Littleton CO	1-800-553-0572
Terratek	Dallas TX	1-800-338-3182
Tierra Madre Environmental Corp.	Ada OK	1-800-377-5129
Truespin Auger Marketing	Corona CA	1-800-642-8437
Westinghouse Groundwater Recovery	Stone Mountain GA	1-800-922-9497
Wright, R.E. Associates Inc.	Middletown PA	1-800-944-6778
Wyo-Ben Inc.	Billings MT	1-800-548-7055
Zimpro Environmental Inc.	Rothschild WI	1-800-826-1476

GEOPHYSICAL APPARATUS & SUPPLIES

Century Geophysical Corp	Las Vegas NV	1-800-722-2014
Geophysics Group Inc The	Escondido CA	1-800-258-1317
Gisco Geophysical Geological Instrument	Denver CO	1-800-523-1526
Slope Indicator	Seattle WA	1-800-426-1791
Sonic Surveys Inc	Mont Belvieu TX	1-800-437-2083
Zonge Engineering	Tucson AZ	1-800-523-9913

GOVERNMENT

Federal Agencies:

National Technical Information Service

	Springfield VA	1-800-336-4700
		1-800-553-NTIS

Peace Corps   Washington DC

	Staff Positions	1-800-424-8580 ext 225
	Overseas Assignments	1-800-424-8580 ext 93

U S Department of Agriculture

Meat and Poultry Food Safety Hotline	1-800-535-4555
Soil Conservation Service	1-800-THE-SOIL

U S Department of the Army

Army Material Command Southern Region FPO	1-800-223-7280

U S Department of Defense

	Defense Mapping Agency	St Louis MO
	Washington DC Area	1-800DMAJOBS
	Outside DC	1-800-777-6104

U S Department of Energy

Civilian Radioactive Waste Managem	Washington DC	1-800-225-6972
Human Experimentation Hotline		1-800-493-2998
Occupational Safety & Health (OSH) Standards Interpretations Response Line		
		1-800-292-8061
Richland Operations Office	Richland, WA	1-800-695-4363

U S Department of Interior

Fish and Wildlife Service	Annapolis MD	1-800-448-8322
Environmental Police Hotline	Boston MA	1-800-632-8075
Geological Survey	Reston VA	1-800USAMAPS

U S Environmental Protection Agency

Alternative Treatment Technology	Rockville MD	1-800-424-9346
Asbestos Ombudsman Clearinghouse	Washington DC	1-800-368-5888
Center for Hazardous Materials		1-800-334-2467
Headquarters Recruitment Center	Washington DC	1-800-338-1350
Indoor Air Quality Information Clearinghouse		1-800-438-4318
Emergency Planning and Community Right-to-Know Information Hotline		
	Washington DC	1-800-535-0202
	(TDD)	1-800-553-7672
EPA Environmental Databases:		
Integrated Risk Information	SSan Jose CA	1-800-872-7654
Federal Endangered Species Protect	Lubbock TX	1-800-447-3813
General Information (CO, MO, ND, SD, UT, WY)		1-800-759-4372
General Information/Publications	Chicago IL	1-800-621-8431
	Chicago IL	1-800-572-2515
Hazardous Waste Management Division	Denver CO	1-800-866-3989
Hazardous Waste Ombudsman	Washington DC	1-800-262-7937
Inspector General Hotline	Washington DC	1-800-424-4000
Mobile Sources (New England states only)		1-800-821-1237
		1-800-631-2700
National Lead Information Center Clearinghouse		1-800-424-5323
National Lead Information Center Hotline		1-800-532-3394
National Pesticides Telecommunictions Network		1-800-858-PEST
National Radon Hotline	Alexandria VA	1800SOSRADON
National Response Center	Washington DC	1-800-424-8802
National Small Flows Clearing Hous	Morgantown WV	1-800-624-8301
Oil and Hazardous Material Technical Assistance		
	Data System Baltimore MD	1-800-CIS-USER
RCRA Superfund Hotline	Washington DC (TDD)	1-800-553-7672
Recycled Products Information Clearinghouse		1-800-424-9346
Region 1 Consumer Product Safety Commission		1-800-638-2772
Region 1 Pesticides Clearinghouse Hotline		1-800-858-7378
Region 2 Hotline (Within region)		1-800-346-5009
Region 3 General Information Hotline		1-800-438-2474
Region 5 Public Affairs Hotline		1-800-621-8431
Region 7 Regional Action Hotline		1-800-223-0425
Region 8 Emergency Response Hotline		1-800-424-8802
Region 8 Toll-Free Line		1-800-227-8917
Region 10 Public Information Center (within region)		1-800-424-4EPA
Safe Drinking Water Hotline	Washington DC	1-800-426-4791
Small Business Ombudsman Clearingh	Washington DC	1-800-368-5888
Solid Waste Information Clearingho	Silver Spring MD	1-800-67SWICH

Storage and Retrieval of US Waterways Parametric Data (STORET)

	User Support Line	1-800-424-9067
Superfund - General Headquarters		1-800-424-9065
Superfund - Community Relations Information		1-800-231-3075
Wastewater Treatment and Information Exchange		
	Bulletin Board System Morgantown WV (BBS)	
		1-800-544-1936
Wetlands Protection	Arlington VA	1-800-832-7828
Whistleblower Hotline - OIG		1-800-424-4000

U.S. Nuclear Regulatory Commission
Recruitment Staff	Washington DC	1-800-368-5642

National Oceanic and Atmosheric Administration (NOAA)
Meteorologists Recruitment	Norfolk VA	1-800-537-4101

State Fish and Game Agencies - Poaching Hotlines:

The National Anti-Poaching Foundation Routing System Number is in Colorado Springs CO. You can place a call from anywhere and be instantly routed to the appropriate state fish & game agency by calling 1-800-800-WARDEN. Within a state, call:

ALABAMA	Game Watch	1-800-272-4263
ALASKA	Fish and Wildlife Safegual	1-800-478-3377
ARIZONA	Operation Game Thief	1-800-353-0700
ARKANSAS	Sport	1-800-482-9262
CALIFORNIA	Cal Tip	1-800-952-5400
COLORADO	Operation Game Thief	1-800-332-4155
CONNECTICUT	Turn In Poachers	1-800-842-4357
DELAWARE	Operation Game Thief	1-800-292-3030
FLORIDA	Wildlife Alert	
	Northwest	1-800-343-1070
	Northeast	1-800-342-8105
	Central	1-800-342-9620
	Everglades	1-800-432-2046
GEORGIA	Turn In Poachers	1-800-241-4113
HAWAII	Conservation Hot Line	1-800-587-0077
IDAHO	Citizens Against Poaching	1-800-632-5999
ILLINOIS	Turn In Poachers	1-800-252-0163
INDIANA	Turn In Poachers	1-800-847-4387
IOWA	Turn In Poachers	1-800-532-2020
KANSAS	Operation Game Thief	1-800-228-4263
KENTUCKY	Report A Poacher	1-800-252-5378
	Natural Resources Department	
	Frankfort KY	1-800-247-4659
LOUISIANA	Help Stop Poaching	1-800-442-2511
MAINE	Operation Game Thief	1-800-253-7887
	Fisheries and Wildlife	
	Augusta ME	1-800-322-3606
MARYLAND	Catch a Poacher	1-800-635-6124
MASSACHUSETTS	Fish and Wildlife Departm	1-800-632-8075
MICHIGAN	Turn In Poachers	1-800-292-7800
MINNESOTA	Turn In Poachers	1-800-652-9093
MISSISSIPPI	Be Smart	1-800-237-6278
MISSOURI	Operation Game Thief	1-800-392-1111
MONTANA	Turn In Poachers	1-800-847-6668
NEBRASKA	Operation Game Thief	1-800-742-7627
NEVADA	Operation Game Thief	1-800-992-3030

NEW HAMPSHIRE	Operation Game Thief	1-800-344-4262
NEW JERSEY	Operation Game Thief	1-800-222-0458
NEW MEXICO	Operation Game Thief	1-800-843-4263
New York	Turn In Poachers & Pollut	1-800-817-7332
NORTH CAROLINA	Wildlife Watch	1-800-662-7137
NORTH DAKOTA	Report All Poachers	1-800-472-2121
OHIO	Turn In a Poacher	1-800-762-2437
OKLAHOMA	Operation Game Thief	1-800-552-8039
OREGON	Turn In Poachers	1-800-452-7888
PENNSYLVANIA Sportsmen Policing Our Ranks Together		
	Northwest	1-800-533-6764
	Northcentral	1-800-422-7551
	Northeast	1-800-228-0789
	Southwest	1-800-243-8519
	Southcentral	1-800-422-7554
	Southeast	1-800-228-0791
SOUTH CAROLINA	Operation Game Thief	1-800-922-5431
SOUTH DAKOTA	Turn In Poachers	1-800-592-5522
TENNESSEE	Stop Poaching	1-800-255-8972
	Wildlife Resources Agency	1-800-262-6704
TEXAS	Operation Game Thief	1-800-792-4263
	Texas Parks & Wildlife	1-800-792-1112
UTAH	Help Stop Poaching	1-800-662-3337
VERMONT	Operation Game Thief	1-800-752-5378
VIRGINIA	Wildlife Violation Hotlin	1-800-237-5712
WASHINGTON	Help Stop Poaching	1-800-477-6224
WEST VIRGINIA	Net Game	1-800-638-4263
WISCONSIN	Turn In Poachers	1-800-847-9367
WYOMING	Stop Poaching	1-800-442-4331

State Government Environmental Agencies:

ARIZONA
Governor's Commission on Arizona Enviro	Phoenix AZ	1-800-826-3257
Department of Water Resources	Phoenix AZ	1-800-352-8488

CALIFORNIA
Department of Conservation Division of Sacramento CA		1-800RECYCLE

COLORADO
Energy Conservation Office	Denver CO	1-800-632-6662
Colorado Environmental Control DivisionCommerce City CO		1-800-732-3060

FLORIDA
OSHA Guard	Clearwater FL	1-800-522-9308
South Florida Water Management DistrictW Palm Beach FL		1-800-432-2045
USDA Animal and Plant Inspection	Gainesville FL	1-800-342-0395

GEORGIA
Agriculture Department District 5	Jesup GA	1-800-874-0258

HAWAII
Conservation Hot Line		1-800-587-0077

IDAHO

Department of Employment	Boise ID	1-800-772-2553
Energy Hotline (ID, OR)	Boise ID	1-800-334-7283
Health and Welfare Division of Environm	Boise ID	1-800-828-6338
US Army Corps of Engineers Water Levels	Ashaka ID	1-800-321-3198
US Labor Department Occupational Safety	Boise ID	1-800-482-1370

ILLINOIS

Environmental Hazard Control	Wauconda IL	1-800-543-1499
Milwaukee Asbestos Information Center	Milwaukee WI	1-800-848-3298
US Army Corps of Engineers	Champaign IL	1-800-252-7122
US National Hazardous Materials Informa	Argonne IL	1-800-752-6367

INDIANA

State Police Hazardous Materials Emerge	Indianapolis IN	1-800-523-2226

KANSAS

US Labor Department Occupational Safety	Wichita KS	1-800-362-2896

KENTUCKY

Department of Agriculture Division of L	Frankfort KY	1-800-372-7127
Natural Resources and Environmental Pro	Frankfort KY	1-800-247-4659

MAINE

Environmental Protection Department	Augusta ME	1-800-453-4013
Fisheries and Wildlife	Augusta ME	1-800-322-3606
Jobs Training	Augusta ME	1-800-245-5627
Low Level Radioactive Waste Authority	Augusta ME	1-800-422-4911

MARYLAND

Department of the Environment Sediment	Baltimore MD	1-800-922-8017
Recycling Information		1800IRECYCLE
Unemployment and Training Department	Baltimore MD	1-800-492-6804

MASSACHUSETTS

Fish and Wildlife Department		1-800-632-8075
Public Safety Department	Brookline MA	1-800-682-9229

MICHIGAN

Central Michigan District Health Depart	Mt Pleasant MI	1-800-332-3258
Intertribal Council of Michigan	Sault St Marie MI	1-800-562-4957

MINNESOTA

Hazard Hotline (Out of State)		1-800-228-5635
Technology	Moorhead MN	1-800-626-3497

MISSISSIPPI

Tennessee Tombigbee Waterway Develop		
	Columbus MS	1-800-457-9739

MISSOURI
| US Agriculture Department Forest Servic | Whitehall MT | 1-800-433-9206 |

MONTANA
| Montana Superfund Hotline | Helena MT | 1-800-648-8465 |

NEBRASKA
| US Labor Department Occupational Safety | Omaha NE | 1-800-642-8963 |

NEW HAMPSHIRE
Department of Agriculture	Concord NH	1-800-562-5274
Department of Health and Welfare	Rochester NH	1-800-862-5300
University of New Hampshire Technology	Durham NC	1-800-423-0060

NEW JERSEY
Air Pollution Index		1-800-782-0160
Asbestos Hotline		1-800-624-2376
Radon Hotline		1-800-648-7263
Seashore Conditions		1-800-648-7263

NEW MEXICO
Highway Department	Grants NM	1-800-824-1007
Human Services Department	Sante Fe NM	1-800-432-6217
Occupational Health and Safety Board		1-800-222-6742

New York
Brookhaven National Laboratory	Upton NY	1-800-282-5195
Environmental Conservation Department	Albany NY	1-800-642-7448
Turn In Poachers & Polluters		1-800-817-7332

NORTH CAROLINA
	Washington NC	1-800-338-7804
Division of Marina Fisheries	Elizabeth City NC	1-800-338-7805
	Wilmington NC	1-800-248-4536
Small Business and Technology Developme	Raleigh NC	1-800-258-0862
University of New Hampshire Technology	Durham NC	1-800-423-0060
Wildlife Resources	Raleigh NC	1-800-662-7137

NORTH DAKOTA
| Resource Education Center | Bismarck ND | 1-800-437-8054 |

OHIO
Attorney General Environmental Investig	Columbus OH	1-800-348-3248
Attorney General Consumer Public Action	Columbus OH	1-800-282-0515
Environmental Agency Hotline	Columbus OH	1-800-686-2330
US National Institute Occupational Safe	Cincinnati OH	1-800-356-4674

OKLAHOMA
| Department of Libraries (EPA & State ag) | Oklahoma City OK | 1-800-522-8116 |

OREGON

Economic Development	Salem OR	1-800-233-3306
Oil and Chemical Spills Emergency Manag	Salem OR	1-800-452-0311
Northwest Power Planning Council	Portland OR	1-800-452-2324

PENNSYLVANIA

Department of Health State Health Line	Harrisburg PA	1-800-692-7254
OSHA Consultation Program	Indiana PA	1-800-382-1241

RHODE ISLAND

USDA Agricultural Stabilization and Con	W Warwick RI	1-800-551-5144

TENNESSEE

Emergency Management	Nashville TN	1-800-258-3300
	Nashville TN	1-800-262-3400
Environmental Management	Solid & Hazardous Waste	1-800-237-7018
Superfund/Underground Tank		1-800-251-3479
Wildlife Resources Agency	Crossville TN	1-800-262-6704

TEXAS

Federal Endangered Species Protection P	Lubbock TX	1-800-447-3813
Human Services Department	San Antonio TX	1-800-742-2565
National Research Laboratory Commission	Desoto TX	1-800-228-3972
Occupational Information Service	Austin TX	1-800-822-7526
Texas Environmental Control Division	Dallas TX	1-800-368-8371

VIRGINIA

Health Regulatory Boards	Richmond VA	1-800-533-1560
Transportation Safety Administration	Richmond VA	1-800-533-1892

WASHINGTON

Emergency Management	Olympia WA	1-800-562-6108
Hazardous Substances	Olympia WA	1-800-633-7585
Industrial Safety and Health Info and A	Olympia WA	1-800-423-7233
Litter Hotline		1-800-LITTERS
Recycling Hotline	Olympia WA	1-800RECYCLE
Social and Health Services Department		1-800-782-0624
		1-800-523-2301
Southwest Air Pollution Control Authori	Vancouver WA	1-800-633-0709
Superfund Hotline		1-800-458-0920
Vehicle Emissions		1-800-272-3780
Waste Reduction Technology		1-800-822-9933
Woodsmoke Hotline		1-800-458-0920

WEST VIRGINIA

Department of Natural Resources Hazardous Waste		1-800-642-3074

WYOMING

Game and Fish Department	Green River WY	1-800-843-8096
	Jackson WY	1-800-423-4113
	Sheridan WY	1-800-331-9834

CANADA

Canadian Center for Occupational Health	ON CN	1-800-668-4284

HAZARDOUS MATERIALS & HAZARDOUS WASTE

AR Manufacturing & Sales Inc.	Oklahoma City OK	1-800-537-3870
ARD Environmental	Columbia MD	1-800-969-2734
ARI Environmental Inc.	Wauconda IL	1-800-851-1674
ASAP Technical Services Inc.	Cleveland OH	1-800-969-0808
Abatement Technologies Inc.	Lawrenceville GA	1-800-634-9091
Advanced Environmental Systems	Newark DE	1-800-220-5330
Allen-Bradley Response Center	Cedar Rapids IA	1-800-223-5354
Allied Signal Inc.	Morristown NJ	1-800-626-4974
Amcon Group International	Santee CA	1-800-531-0794
American Norit Company Inc.	Atlanta GA	1-800-641-9245
American Safety & Abatement Products In	St Louis MO	1-800-489-8535
American Sigma	Medina NY	1-800-635-1230
American Sigma	Medina NY	1-800-635-4567
American Tank & Boiler	Quincy MA	1-800-498-9696
American Trucking Association	Alexandria VA	1-800-282-4563
Anderson 2000	Peachtree GA	1-800-241-5424
Ansul Fire Protection	Marinette WI	1-800-346-3626
Associated Design & Manufacturing Co.	Alexandria VA	1-800-837-8257
Aztec Environmental Services Div	Norman OK	1-800-879-0255
BHA Group	Kansas City MO	1-800-821-2222
Baghouse Services Inc.	Aiken SC	1-800-437-1168
BioGee International Inc.	Houston TX	1-800-299-3111
Bio-Nomic Services Inc.	Charlotte NC	1-800-621-4342
Brinecell Inc.	Salt Lake City UT	1-800-468-8241
Burlington Environmental	Seattle WA	1-800-228-7872
CC Lynch & Associates	Pass Christian MS	1-800-451-2252
Carbtrol Corp	Westport CT	1-800-242-1150
Celgene Biotreatment Systems	Warren NJ	1-800-253-1596
ChemFab Protective Clothing	Merrimack NH	1-800-451-6101
Chemical Safety Technology Inc.	San Jose CA	1-800-536-5277
Chemical Waste Management Inc.	Oak Brook IL	1-800-843-3604
Civacon	Cincinatti OH	1-800-526-5657
Clean Air Engineering	Palatine IL	1-800-627-0033
Clean Water Systems Northwest	Sumner WA	1-800-525-7444
Clements Associates Inc.	Newton IA	1-800-247-6630
Construction Minerals Inc.	Buffalo NY	1-800-662-6561
Containment Systems	Cocoa FL	1-800-632-5640
Conwed Bonded Fiber	Riverside NJ	1-800-526-5138
Copp Environmental Products	Pomona CA	1-800-779-2677
Correct Nonwovens Corp	Huntington NY	1-800-234-2012
CURA Inc.	Dallas TX	1-800-486-7117
CYA Products	Long Island NY	1-800-247-9274
DMP Corp	Ft Mill SC	1-800-845-3681
Delphian Corp.	Northvale NJ	1-800-288-3647
Direct Safety Co.	Phoenix AZ	1-800-528-7405
Edwards Engineering Corp	Pompton Plains NJ	1-800-835-3222
Eilco	Concord CA	1-800-648-9355
Encycle/Texas Inc. Sub. of Asarco Inc.	Corpus Christi TX	1-800-443-0144
Enercon Systems	Atlanta GA	1-800-572-1835
EnSys Environmental Products		1-800-242-7472

Enterra Instrumentation Technologies	Exton PA	1-800-634-4046
Environmental Chemical Association	Marlboro NJ	1-800-327-3634
Environmental Container Corp.	Delafield WI	1-800-729-7137
Environmental Contracting	Alsip IL	1-800-331-1945
Environmental Education Enterprises	Columbus OH	1-800-792-0005
Environmental Protection Inc.	Mancelona MI	1-800-345-4637
Environmental Tank Systems Inc.	Springer OK	1-800-374-8463
Enviroscan Corp	Rothschild WI	1-800-338-7226
Erichson Fred M	Metaire LA	1-800-824-8973
Flint Environmental Services	Tulsa OK	1-800-525-7820
Fluid Controls	Huntsville AL	1-800-462-0860
Four Seasons Environmental Inc.	Greensboro NC	1-800-868-2718
Free Enterprise Systems Inc.	Rapid City SD	1-800-894-5041
Geoguard	Medina NY	1-800-645-7654
Geopure Continental Systems and Services		1-800-342-1103
Gibraltar National Corp	Detroit MI	1-800-442-7258
Global Equipment Co.	Port Washington WA	1-800-645-1232
Gore, W.L. & Associates	Phoenix AZ	1-800-438-6981
Gundle Lining Systems Inc.	Houston TX	1-800-435-2008
Guzzler Manufacturing Inc.	Birmingham AL	1-800-822-8785
Halliburton NUS Environmental Corp.	Gaithersburg MD	1-800-368-2755
Hazco Environmental Services	Calgary AB CN	1-800-667-0444
Hazco Services Inc.	Dayton OH	1-800-332-0435
Hazstor	Des Plaines IL	1-800-798-9250
HCL Labels Inc.	Sunnyvale CA	1-800-421-6710
Heritage Environmental Services	Indianaplois IN	1-800-827-0476
Hime Environmental Products Inc.	Albuquerque NM	1-800-735-4463
Horizon Environmental Well Products	Wilmot SD	1-800-832-8369
Hoyt Corp	Westport MA	1-800-343-9411
H2Oil	Hayward CA	1-800-835-3221
ILC Dover (protective clothing)	Fredericka DE	1-800-631-9567
Industrial Environmental Products Inc.	Al Pharetta GA	1-800-233-1001
Industrial Environmental Supply Inc.	Greensboro NC	1-800-768-4817
Industrial Scientific Corp.	Oakdale PA	1-800-338-3287
Innovative Material Systems Inc.	Olathe KS	1-800-800-4010
InVitro International	Irvine CA	1-800-647-6725
ITEX Enterprise Inc.	Dallas TX	1-800-880-9053
J B Hunt Special Commodities	Lowell AR	1-800-643-3622
JMC Soil Investigation Equipment/Clemen	Newton IA	1-800-247-6630
JSB	Marietta GA	1-800-551-1207
Justrite Manufacturing Co.	Des Plaines IL	1-800-798-9250
Keystone Dynamics Inc.	Broomall PA	1-800-356-7613
KISZ Environmental Training Co.	Ashland City TN	1-800-227-5685
KVB Analect FT-IR	Irvine CA	1-800-326-2328
Labelmaster	Chicago IL	1-800-621-5808
LARCO Corp	Newton MA	1-800-654-0308
Leapfrog Technologies	Altoona PA	1-800-443-7647
Liquid Waste Technology Inc.	Somerset WI	1-800-243-1406
MDA Scientific Inc.	Lincolnshire IL	1-800-344-4632
Magid Glove & Safety Mfg Co.	Chicago IL	1-800-444-8010
Max Bac Bioremediation		1-800-695-8085
Mercury Refining Company	Latham NY	1-800-833-3505
Mettam Safety Supply Co.	Danville IL	1-800-252-3009
Midwesco Filter Resources	Winchester VA	1-800-336-7300
Mineral Springs Corp.	Port Washington WI	1-800-932-6216
Monroe Environmental Corp	Monroe MI	1-800-992-7707
NAO Inc.	Philadelphia PA	1-800-523-3495
National Environmental Products	Newfield NJ	1-800-542-6816

National Filter Media	Moorestown NJ	1-800-628-5110
National Seal Co.	Aurora IL	1-800-323-3820
NEO Corp	Waynesville NC	1-800-822-1247
Neotronics	Gainesville GA	1-800-535-0606
NEPCCO Equipment	Ocala FL	1-800-277-3279
New Pig Corp	Tipton PA	1-800-468-4647
Nilfisk of America Inc.	Malvern PA	1-800-645-3845
N-Viro Energy Systems Ltd.	Toldeo OH	1-800-66NVIRO
Oil Pollution Advisory Services	E Rutherford NJ	1-800-438-6727
OSHA - Soft Corp	Amherst NH	1-800-446-3427
PCP Inc.	W Palm Beach FL	1-800-637-5307
PacTec Inc.	Clinton LA	1-800-272-2832
Pacific Environmental Services Co.	Port Townshend WA	1-800-222-9219
Pacific Industrial Service Corp	Long Beach CA	1-800-359-6377
Parker Systems	Norfolk VA	1-800-666-0006
Parkline Inc.	Winfield WV	1-800-786-4855
Pensacola Pollution Control Inc.	Pensacola FL	1-800-642-7445
Permatron Corp	Franklin Park IL	1-800-882-8012
Plastic Fusion Fabricators Inc.	Huntsville AL	1-800-356-1480
Pollution Control Inc.	Florence KY	1-800-446-4724
Pollution Control Industries, Inc.	S Chicago Heights IL	1-800-388-7242
Poly-Flex Inc.	Grand Prairie TX	1-800-527-3322
Poly Hi Solidur	Ft Wayne IN	1-800-628-7264
Poscon	Conroe TX	1-800-437-3037
Precip Tech Inc.		1-800-336-2585
Professional Environmental Trainers Association		1-800-788-7464
Quadrel Services Inc.	Ijamsville MD	1-800-878-5510
Quantix /Agri-Diagnostics Environmental	Cinnaminson NJ	1-800-322-5487
QED Environmental	Ann Arbor MI	1-800-343-2026
Quest technologies	Oconomowoc WI	1-800-245-0779
RECRA Environmental Inc.	Amherst NY	1-800-52RECRA
REECO	Morris Plains NJ	1-800-486-1507
Rollins Environmental Services	Wilmington DE	1-800-992-7837
Ross Incineration Services Inc.	Grafton OH	1-800-878-7677
Ross Transportation Services Inc.	Grafton OH	1-800-783-6555
Rust Environmental	Sheboygan WI	1-800-242-7601
Rust Environmental & Infrastructure	Greenville SC	1-800-868-0373
Rust International Corp.	Birmingham AL	1-800-247-3122
S & G Enterprises Inc.	Milwaukee WI	1-800-233-3721
Safe-T-Tank Corp	Meridian CT	1-800-545-1865
Safe Environmental Storage Corp.	Bensenville IL	1-800-593-7233
Safety Storage Inc.	Hollister CA	1-800-344-6539
Sahlberg Equipment	Seattle WA	1-800-825-1808
Salem Optical Co. Inc.	Winston-Salem NC	1-800-642-0493
Sanders Lead Company	Troy AL	1-800-633-8744
Scientific Specialties Service	Randallstown MD	1-800-648-7800
SEC Donahue/Rust International	Greenville SC	1-800-868-0373
Sengineering	Novato CA	1-800-822-3459
Sensidyne Inc.	Clearwater FL	1-800-451-9444
Shield's Manufacturing Co Inc.	Oxnard CA	1-800-552-8783
Smith Engineering Company	Ontario CA	1-800-959-5732
Soil Remediation Co.	Denver CO	1-800-441-1968
Sonoco Fiber Drum	St Louis MO	1-800-548-2482
Southdown Inc.	Fairborn IL	1-800-762-0040
Southern Environmental	Pensacola FL	1-800-282-8734
Spill Management Inc.	Traverse City MI	1-800-800-6084
Sprung Instant Structures		1-800-528-9899
Staplex Co	Brooklyn NY	1-800-221-0822

Staclean Diffuser	Salisbury NC	1-800-438-3850
Super Products Co.	Milwaukee WI	1-800-837-9711
Surface Combustion Inc.	Maumee OH	1-800-537-8980
Sur-Lite Corp	Sante Fe Springs CA	1-800-432-8818
Tanknology	Agawam MA	1-800-666-2605
Telpac Tower Packing	Boston MA	1-800-523-0948
Therma Fab Inc.	Lexington SC	1-800-762-1712
3M Ceramic Materials		1-800-328-1684
Tier Remediation Inc.	New Castle DE	1-800-422-0308
TN Technologies Inc.	Round Rock TX	1-800-736-0801
Tonto Drilling Services Inc.	Fontana CA	1-800-350-6611
Tracer Research Corp	Tucson AZ	1-800-394-9929
Triangle Environmental Services	Research Triangle Park NC	1-800-367-4862
Unocal Corp	Fullerton CA	1-800-323-8647
US Ecology	Louisville KY	1-800-999-7160
Wahlco	Leawood KS	1-800-228-5910
Watersaver Denver CO	Commerce City CO	1-800-525-2424
West Hazmat Corp	Englewood CO	1-800-821-8428
Wheelabrator Air Pollution Control	Pittsburgh PA	1-800-327-8727
Wheelabrator Air Pollution Control		1-800-458-8522
Williams & Associates	Clearwater FL	1-800-277-9802
Witt, F.C. Associates	Claremore OK	1-800-323-3335
Otto York Co. Inc.	Parsippany NJ	1-800-524-1543
Zimpro Passavant	Avon Lake OH	1-800-421-3144
Zimpro Passavant	Irondale AL	1-800-633-9501
Zimpro Passavant	Rothschild WI	1-800-826-1476

INCINERATORS

Sharps Incinerator Of Fort Atkinson Inc	Ft Atkinson WI	1-800-673-9247
Therm Tec	Elyria OH	1-800-875-3614

INDUSTRIAL CLEANING & SERVICES

4 Seasons Industrial Services	Nashville TN	1-800-848-8720
Allied Hydro Blasters	Tampa FL	1-800-382-4128
Alpheus Cleaning Technologies	Rancho Cucamonga CA	1-800-445-6131
Benton Harbor Laundry	Benton Harbor MI	1-800-664-4568
Brand Utility Cleaning Services	Rockdale TX	1-800-344-6827
C & K Industrial Services	Cleves OH	1-800-852-9468
CEDA	Bellingham WA	1-800-824-2332
Caligo Ltd	Sunrise Beach MO	1-800-654-2880
Care-N-Clean Drema Kromer	Punxsutawney PA	1-800-831-4862
Clarke Industries Inc	Bowling Green OH	1-800-331-7692
Deep South Industrial Cleaning Inc	Jesup GA	1-800-462-8905
Eldorado Chemicals	San Antonio TX	1-800-292-5372
Eldorado Chemicals	San Antonio TX	1-800-531-1088
Electrosolve Services	Costa Mesa CA	1-800-247-0505
Glic Environmental	Toledo OH	1-800-338-4542
Halliburton Industrial Services	Citronelle AL	1-800-932-5326
Hotsy Co	Hyde Park MA	1-800-544-7790
Houston Environmental Solutions	Deer Park TX	1-800-262-5326
Hunt Cleaners Inc.	Cozad NE	1-800-262-4568
Industrial Products	Knoxville TN	1-800-835-8983
Insta Clean		1-800-331-6405
Jam Enterprises	Michigan City IN	1-800-553-1136

Jet Blast Inc	Hopewell VA	1-800-538-2527
Leak Repair	Brunswick GA	1-800-225-0443
Mapps Cleaning Supplies	Culver City CA	1-800-266-7627
Miller Contracting Co Inc	Dallastown PA	1-800-233-1992
Mole Master Services Corp.	Marietta OH	1-800-322-6653
Municipal Industrial Vacuum	Vancouver WA	1-800-453-9927
National Clean Ceiling	W Chester PA	1-800-525-3265
National Industrial Maintenance	Dearborn MI	1-800-952-0111
National Plant Services	Long Beach CA	1-800-445-3614
Panther Chemical	Ft Worth TX	1-800-433-7664
Phoenix Labs	Fallsington PA	1-800-521-4469
RW Evans And Associates		1-800-336-1614
S & S Enterprise	Freeburg IL	1-800-245-5326
Singer Management-Conklin Distributors	Liberty MO	1-800-528-2189
Skasol	Pagedale MO	1-800-752-7657
Skill Clean	Napa CA	1-800-356-2532
Steamatic Of Tallahassee Inc	Tallahassee FL	1-800-255-7832
TNT Cleaning Services	Raleigh NC	1-800-865-4759
Turco Purex	Marion OH	1-800-848-0085
US Polychemical Corp	Spring Valley NY	1-800-431-2072
Vac N Jet Environmental	Pueblo CO	1-800-358-3767
Video Industrial Services	Bham AL	1-800-826-3498
Wachs Technical Services	Charlotte NC	1-800-841-1302
Westinghouse Industry Services-	Pittsburgh PA	1-800-441-3134

INDUSTRIAL CONSULTANTS

D & J Carpet Floor Cleaning Service	Albemarle NC	1-800-662-0674
EMJ Industrial Supply	Oceanside NY	1-800-227-7156
Eckert Industrial Supply	Jasper IN	1-800-842-6933
G P S Technologies	Baton Rouge LA	1-800-264-2436
Hagerstown Industrial Supply	Hagerstown MD	1-800-733-6111
IBM Industrial Consulting Services Info	Boca Raton FL	1-800-472-7463
IMCS	Cincinnati OH	1-800-638-3132
Kasper Consultants Inc	Elyria OH	1-800-835-9952
Midwest Institute For Ethical Conduct	Manhattan KS	1-800-682-6731
Prevention Plus	Minneapolis MN	1-800-887-2282
Quail Enterprises	Gary TX	1-800-688-8921
Straightline Optical Service Inc	Cumming GA	1-800-638-8936
Studio Tech Inc	Newport Beach CA	1-800-887-8834
Team Development Group	Bloomington MN	1-800-628-1238
Technology Strategic Planning	Stuart FL	1-800-677-0335
Vibration Solutions Co	W Des Moines IA	1-800-238-6762

INSECTICIDES

AAAW Guaranty Pest Elimination Inc	Thompson CT	1-800-553-5528
Best Enterprises Ltd.	Winterville NC	1-800-762-2457
Better Crops	W Chester IA	1-800-628-6742
Blue Diamond Exterminating Co Inc	Hyden KY	1-800-322-6732
Borden Pest Control	North Augusta SC	1-800-782-4433
Clarke Mosquito Control Prod & Serv	Las Vegas NV	1-800-323-5727
Coulston Products	Easton PA	1-800-445-9927
Environmentally Safe Products	St Joe TX	1-800-995-7134
Epico Inc	San Antonio TX	1-800-532-5808

Expert Pest Control	Leominster MA	1-800-235-3093
FMC Specialty Products	Princeton NJ	1-800-528-8873
Green Solutions Intl Inc	Coral Gables FL	1-800-222-7280
Impact Pest Management	Tarpon Springs FL	1-800-344-9190
Jackson's Pest Control	Dillon SC	1-800-622-2671
Jaxon Inc	Raymond MS	1-800-548-0035
Legend Pest Elimination	Hampton VA	1-800-457-2993
McCloud Pest Control-	Girard IL	1-800-752-6750
McCloud Pest Control	Girard IL	1-800-222-4798
McCloud Pest Control	Kansas City MO	1-800-231-8660
McCloud Pest Control	Indianapolis IN	1-800-352-2315
McCloud Pest Control	Girard IL	1-800-421-6437
McCloud Pest Control	Decatur IL	1-800-428-0026
McCloud Pest Control	Hazelwood MO	1-800-733-5735
Metro Termite & Pest Control	North Augusta SC	1-800-524-9335
Mike Doty's Termite & Pest Control	Warsaw MO	1-800-525-3801
Mountain Pest Control	Mt Pocono PA	1-800-638-7820
Natural Insecto GHP	Longview TX	1-800-437-9068
Organic Plus Of South Texas	Adkins TX	1-800-417-2261
Organic Plus	Albuquerque NM	1-800-933-2278
PF Harris	Jacksonville FL	1-800-637-0317
Pest Control Supplies	St Louis MO	1-800-821-5689
Pest Control Supplies	St Louis MO	1-800-892-7177
Presidential Pest Control	N Conway NH	1-800-649-6508
Professional Pest Control Products Of Florida	Tampa FL	1-800-232-4271
Rentokil Pest Control	Norcross GA	1-800-732-3716
Texas Exterminating	Plainview TX	1-800-831-3368
Truly Nolan Of America Exterm Inc	Tucson AZ	1-800-528-3442
VecTec Inc	Orlando FL	1-800-367-1299
Vulcan Pest Control	W Palm Beach FL	1-800-346-0971
Western Termite & Pest Control	Parsippany NJ	1-800-544-2847

INSPECTION SERVICES

AAA Home Inspection	Port St Lucie FL	1-800-337-9540
American Builders Consultants	Grand Ledge MI	1-800-368-4734
American Building Inspections Inc	Willoughby OH	1-800-851-1655
Amerispec	Orange CA	1-800-628-9104
Amerispec	Orange CA	1-800-631-4113
Amview	New Paltz NY	1-800-851-2420
Asset Control Services Inc	Woburn MA	1-800-445-1309
Atlantic Home Inspection	Hilton Head Island SC	1-800-548-3551
Bryant's Inspection Bureau	Logansport LA	1-800-671-4623
Building Inspection Services	Miami FL	1-800-255-3317
Building Inspector Of America	Wakefield MA	1-800-321-4677
Cantech Inspections Ltd	Bellingham WA	1-800-289-9101
Central Electric Inspection Bureau	Youngstown OH	1-800-437-0932
Certified Measurement	Centerville GA	1-800-525-0408
Checkpoint	Canton MA	1-800-828-8377
Coastal Inspection Services	Chelsea MA	1-800-231-0205
Comtrol Services	Pasadena TX	1-800-232-1884
Comtrol Services	Pasadena TX	1-800-552-5775
Cresti	River Edge NJ	1-800-441-2578
Curleys Inspection Services	Monahans TX	1-800-462-1219
DCI Calibration	Searcy AR	1-800-443-0247
Diversified Fleet Services	Houston TX	1-800-643-3033
Diversified Inspections, Inc.	Phoenix AZ	1-800-992-1111

FCS Inspection Services	Lyman SC	1-800-635-0746
Faulkner Entresource	Houston TX	1-800-929-3002
Freestate Home Inspection Service	Annapolis MD	1-800-619-0460
GKS Inspection Services Hotline	Livonia MI	1-800-346-3898
Hartford Steam Boiler	Pittsburgh PA	1-800-444-4724
Heritage Inspection Service	Louisville KY	1-800-242-2294
Home Inspection Services Of Michigan	Alto MI	1-800-637-4503
Home Sweet Home Inspection Services	Walterboro SC	1-800-722-7895
Homepro Inspection Service Inc	Tacoma WA	1-800-423-0114
Homewood Inspections	Hamburg MI	1-800-540-3263
Homexam	Marysville OH	1-800-644-4974
House Master Of America	Bound Brook NJ	1-800-323-4677
Indiana Home Inspections	Valparaiso IN	1-800-726-0813
Inspection Services Associates	Miramar FL	1-800-582-9659
Inspection Specialists	Midland TX	1-800-286-6142
Inspections Unlimited	Independence OR	1-800-235-6790
Intertek Services Corp	Fairfax VA	1-800-336-0151
Intertek Services Corp	Rolling Hills Estates CA	1-800-421-0369
J M I	Parma OH	1-800-772-0436
Kennedy Coating Inspection Service	Cypress TX	1-800-882-5344
LDAR Services	Corpus Christi TX	1-800-356-5327
M L K Inspections	Valencia CA	1-800-655-2510
M Q S Inspection Inc	Santa Fe Springs CA	1-800-528-0107
MDIA	Rome NY	1-800-547-6342
MTI Inspection Services	Bedford TX	1-800-541-1462
McDermott Energy Services Inc	Houston TX	1-800-233-0284
Middle Department Inspection	Wexford PA	1-800-992-6342
Midwest Inspection Service Inc	Eden Prairie MN	1-800-392-3570
Mike Hean's Pipeline Service	Stockton CA	1-800-238-9751
National Fields Representative	Claremont NH	1-800-554-0937
National Home Buyers Services	Cooper City FL	1-800-799-6427
National Property Inspections	Prescott AZ	1-800-452-6997
Oldham Brian	Lansing MI	1-800-362-4251
PTS Intl	Brenham TX	1-800-767-7170
Parra Building Consultants	San Diego CA	1-800-222-1910
Perfection Probes	Palatine IL	1-800-521-8868
Phares Kenneth G	Seven Hills OH	1-800-494-0030
Precision Measuring Corp-	Fraser MI	1-800-243-2520
Pro Q C Systems	New York NY	1-800-362-3639
Q S L Inc	Hunt Valley PA	1-800-253-2984
R K Inspection Service	Saucier MS	1-800-289-2251
SGS Industrial Services	San Leandro CA	1-800-638-4677
Sierra Home Service	San Ramon CA	1-800-924-4204
Sierra Home Service	San Ramon CA	1-800-924-7114
Spec Consultants	Lynchburg VA	1-800-747-7732
Spec Consultants	Trafford PA	1-800-837-7732
Steward Services	Ft Worth TX	1-800-397-5317
Sutton Inspection Bureau Inc	Paramus NJ	1-800-858-9337
TEI Analytical Services Inc	Washington PA	1-800-542-0399
Texas NDE Technologies	Katy TX	1-800-772-3044
Turner Of The Century	Windsor VT	1-800-638-4560
Unicon Intl Limited	Lyons IL	1-800-358-7978
Vetco Pipeline Services Inc	Houston TX	1-800-231-1725
Warren Wolf Associates Inc	Dover DE	1-800-824-6022
Wilson Johnnie Inspections	Angleton TX	1-800-578-3441

JOURNALS & PUBLICATIONS

ABSEARCH Moscow ID		1-800-867-1877
American Journal of Botany	Lawrence KS	1-800-627-0932
American Water Works Association	Denver CO	1-800-926-7337
Asbestos Victims Special Fund Trust	Philadelphia PA	1-800-447-7590
Bottle-Can Recycling Update	Portland OR	1-800-227-1424
Business Publishers Inc.	Silver Spring MD	1-800-274-0122
Cole-Parmer Instrument	Niles IL	1-800-323-4340
CRC Press/Lewis Publishers Inc.	Boca Raton FL	1-800-272-7737
DIRECT CONTACT PUBLISHING	Kennewick WA	1-800-457-8746
F W Dodge	New York NY	1-800-541-9913
Earth Journal		1-800-825-0061
Earth Journal - Editors Publications		1-800-333-8857
Environment Magazine		1-800-365-9753
Environmental Resource Center	Cary NC	1-800-537-2372
Environment Today	Marietta GA	1-800-966-3976
Genium Publishing Corp.	Schenectedy NY	1-800-243-6486
The Genuine Article	Philadelphia PA	1-800-523-1850
Institute for Scientific Information	Philadelphia PA	1-800-336-4474
JJ Keller & Associates	Neenah WI	1-800-558-5011
Labelmaster	Chicago IL	1-800-621-5808
Lab Safety Supply	Janesville WI	1-800-356-0783
Oxford University Press	New York NY	1-800-451-7556
Pollution Engineering Locator	Des Plaines IL	1-800-662-7776
Resources Magazine of Environmental Management		1-800-544-3117
Restek Corp	Bellefonte PA	1-800-356-1688
Restoration Ecology (Blackwell Scientific Publications)		
	Cambridge MA	1-800-215-1000
United Nations Publications	New York NY	1-800-253-9646
University of California Press		1-800-822-6657
US Water News Inc.	Halstead KS	1-800-251-0046
Van Nostrand Reinhold	New York NY	1-800-842-3636

LABORATORIES - ANALYTICAL, ENVIRONMENTAL, HAZARDOUS, RADIOLOGICAL & TESTING

Aaron Environmental Specialists	Waterbury CT	1-800-248-9858
ABC Laboratories	Columbia MO	1-800-533-0222
ADI Systems Inc.	ON CN	1-800-561-BVF1
A & A Lab	Springdale AR	1-800-962-7120
A & E Testing	Baton Rouge LA	1-800-673-5530
A & L Agricultural Labs of Memphis Inc.	Memphis TN	1-800-624-2044
A & L Analytical Laboratories Inc.	Memphis TN	1-800-264-4522
Accredited Laboratories Inc.	Carteret NJ	1-800-254-5227
Accuchem Laboratories	Richardson TX	1-800-451-0116
Accuract Testing Plus	Houston TX	1-800-446-0692
Addison Biological Laboratory	Fayette MO	1-800-331-2530
Aegis Analytical Laboratories	Nashville TN	1-800-533-7052
Aeron Biotechnology	San Leandro CA	1-800-367-3296
AGP Lab Inc.	Arlington TX	1-800-247-6030

Allied Clinical Laboratories	Tucson AZ	1-800-851-2404
Alpha Analytical Labs	Westborough MA	1-800-624-9220
Alpha Chemical and Biochemical Labs	Petaluma CA	1-800-221-8607
Alpha Resources	Stevensville MI	1-800-833-3083
Ameri-Tech Diagnostics	Ivyland PA	1-800-356-6033
Alpha Resources Inc.		1-800-833-3083
American Analytical Labs	Tucson AZ	1-800-658-8601
American Analytical Labs	Tucson AZ	1-800-279-6181
American Bio Products	Parsippany NJ	1-800-242-7360
American Biotechnologies	Cambridge MA	1-800-542-2281
American Environmental Laboratories	Leominster MA	1-800-522-0094
American Laboratory	Frankfort IL	1-800-533-2539
American Medical Laboratories	Fairfax VA	1-800-336-3718
American Standards Testing Bureau Inc.	New York NY	1-800-221-5170
American Testing and Research	Cleveland OH	1-800-322-3371
American Testing Labs Inc.	Salt Lake City UT	1-800-847-2293
American Testing Lab	S Salt Lake City UT	1-800-237-7166
American Waste Processing Ltd.	Maywood IL	1-800-841-6900
Ameritech Labs	Bayside NY	1-800-451-1117
Amitech	Omaha NE	1-800-752-4334
ANA Lab	Bellmawr NJ	1-800-648-2625
Anachem	Dallas TX	1-800-942-6891
Analyst Inc.	Torrance CA	1-800-336-3637
Analysts Inc.	Oakland CA	1-800-424-0099
Analysts Services	Stafford TX	1-800-248-7776
Analytical Biosystems Corp	Warwick RI	1-800-262-6520
Analytical Chemical Services of Columbi	Columbia MO	1-800-842-2272
Analytical Technologies Inc.	Ft Collins CO	1-800-443-1511
Analytical Testing Consultants, Inc.	Kannapolis NC	1-800-733-3193
Analytikem Inc.	Rock Hill SC	1-800-444-2536
Anatomical Pathology Services	Parsons KS	1-800-572-4277
Ani Lytics	Gaithersburg MD	1-800-237-2815
Animal Reference Lab	Houston TX	1-800-346-5493
Apollo Environmental	Gibsonton FL	1-800-348-3181
Applied Computer Systems	Durham NC	1-800-932-9356
Applied Environmental Science Inc.		1-800-626-8089
Aqua Tech	Metamora OH	1-800-858-8869
Aqua Tech Environmental Consultants	Marion OH	1-800-783-5991
Aqua Tech Environmental Consultants Inc	Sanford NC	1-800-522-2832
Aqua Tech Environmental Laboratories In	Marion OH	1-800-783-5991
Aqua Test Inc.	Maple Valley WA	1-800-221-3159
Aquatic Bioassay Lab	Baton Rouge LA	1-800-227-7671
Aquatic Biosystems Inc.		1-800-331-5916
ARDL Lab	Mt Vernon IL	1-800-842-7134
Arm Medical Lab	Clearwater FL	1-800-362-9152
Arthur Technology	Fond Du Lac WI	1-800-328-7518
Arts Manufacturing and Supply	American Falls ID	1-800-635-7330
ASAP Technical Services Inc.	Cleveland OH	1-800-969-0808
Assay Technology Inc.		1-800-833-1258
ASTB/Analytical Services Inc.		1-800-221-5170
AT Lab	Grand Prairie TX	1-800-522-2857
ATC Environmental Inc	Sioux Falls SD	1-800-522-9675
Athena Analytical Laboratory	Chicago IL	1-800-578-8030
Atlantic Environmental Lab	W Haven CT	1-800-341-5588
Atlas Testing Lab	Los Angeles CA	1-800-532-8810
Averill Environmental Laboratory Inc.	Plainville CT	1-800-206-1426
Azimuth Inc.	Charleston SC	1-800-537-0336
Averill Environmental Laboratory Inc.	Plainville CT	1-800-206-1426

Barnett Laboratories	New Orleans LA	1-800-542-4004
Barringer Laboratories	Mississauga ON CN	1-800-263-9040
Bay West Inc.		1-800-279-0456
BEC Laboratories Inc.		1-800-331-6561
Bechtol Engineering and Testing	Orange City FL	1-800-526-4061
Belmonte Park Environmental Laboratorie	Dayton OH	1-800-818-5227
Berkshire Analytical	Kinderhook NY	1-800-348-9840
Bio-Technics Lab Inc.	Los Angeles CA	1-800-331-9958
Bioanalytical Technologies	Chicago IL	1-800-872-5221
Bioproducts for Medicine (allergy testi	Tempe AZ	1-800-426-6785
Bioscience Inc.		1-800-627-3069
Biosystems Inc.	Loganville GA	1-800-346-2467
Biotransformations	Colorado Springs CO	1-800-443-9628
Biotransformations Inc.	Colorado Springs CO	1-800-642-8926
BPL Toxicology Laboratory	Tarzana CA	1-800-492-0800
Brooks Lab	Weston CT	1-800-843-1631
Brown and Caldwell Consultants	Pleasant Hill CA	1-800-727-2224
Brown & Root Environmental	Houston TX	1-800-368-2755
Bryson Industrial Services	Lexington SC	1-800-359-7027
Buck Scientific Inc.	Piscataway NJ	1-800-531-2436
CAE Analytical Services	Palatine IL	1-800-627-0033
Calibrations	Latham NY	1-800-828-3818
Calscience Environmental Labs	Sante Fe Springs CA	1-800-383-6363
Cambridge Laboratory	Worthington MN	1-800-635-5086
Caprock Lab	Midland TX	1-800-451-1172
CAVL Inc.	Amarillo TX	1-800-262-2486
CDS Lab	Durango CO	1-800-553-6266
Cedar River Lab	Mason City IA	1-800-323-4858
Cenref Labs	Brighton CO	1-800-634-0497
		1-800-831-2791
Centech Lab	St Louis MO	1-800-423-6832
Center for Applied Engineering Inc.	St Petersburg FL	1-800-637-3225
Central Florida Testing Labs	Largo FL	1-800-248-2385
Cenvet Laboratory	Woodstein NY	1-800-423-6838
Certified Engineering and Testing	Weymouth MA	1-800-443-4353
Certified Testing Lab	Bordentown NJ	1-800-648-7727
Charles River Laboratories	Wilmington MA	1-800-522-7287
Charles River Professional Services	Wilmington MA	1-800-338-9680
Chematox Lab	Boulder CO	1-800-334-1685
Chemetrics	Calverton VA	1-800-356-3072
Chemical Research Labs	Garden Grove CA	1-800-522-1275
Chemir/Polytech Laboratories Inc.	St Louis MO	1-800-659-7659
Chemron	San Antonio TX	1-800-572-6955
ChemServe Environmental Analysis	Milford NH	1-800-675-1868
Chemwest Lab	Sacramento CA	1-800-562-6550
Clamp-Central Laboratory	Timonium MD	1-800-331-2558
Clean Air Engineering	Palatine IL	1-800-627-0033
Clinical Lab of San Bernadino	San Bernadino CA	1-800-541-8905
Columbia Laboratory Services	Monroe NC	1-800-952-2181
Coast-To-Coast Analytical Services Inc.	Valparsiso IN	1-800-678-6522
Coast-To-Coast Analytical Services Inc.	Camarillo CA	1-800-233-2088
Compuchem Laboratories	Research Triangle Park NC	1-800-833-5097
Conam Inspection	Itasca IL	1-800-826-5638
Conam Inspection	Itasca IL	1-800-982-6626
Conam Inspection Inc.	Gary IN	1-800-448-3780
Conam Inspection	Itasca IL	1-800-828-4694
Connecticut Testing Laboratories Inc.	Meriden CT	1-800-394-9151
Consolidated NDE Inc.	Woodbridge NJ	1-800-451-6069

Container Testing Laboratory		1-800-221-5170
Continental Analytical Services	Salina KS	1-800-535-3076
Controls for Environmental Pollution In	Sante Fe NM	1-800-545-2188
CQS	Harrisburg PA	1-800-762-5837
Craven Lab Inc.	Austin TX	1-800-531-5092
Curtin Matheson Scientific	Federal Way WA	1-800-323-3987
Curtis and Thompkins	Los Angeles CA	1-800-522-1878
DCI Calibration	Madison AL	1-800-328-4834
Daigger A & Co	Wheeling IL	1-800-621-7193
Daily Analytical Laboratory	Peoria IL	1-800-752-6651
Data Chem Lab	Richland WA	1-800-643-3924
Data medical Associates Inc.	Arlington TX	1-800-433-7224
Datachem	Cincinnati OH	1-800-458-1493
DCM Science Lab	Lakewood CO	1-800-852-7340
Dellavalle Lab Inc.	Fresno CA	1-800-228-9896
Delmarva Lab	Salisbury MD	1-800-637-5227
Delta Mobile Testing Inc.	Buckner KY	1-800-321-7504
Dental Laboratory Discount Supply	Branford CT	1-800-243-4571
Digicolor Inc.		1-800-848-6448
Diversified Analytical Services	Inglewood CA	1-800-862-9310
Diversified Inspections	Phoenix AZ	1-800-992-1111
Doe Valley Farms Inc.	Bentonville AR	1-800-662-8669
Drew Industrial Division - Ashland Chemical		1-800-526-1015
Drug Testing Consortium	Eunice LA	1-800-447-2974
Dset Lab Inc.	Phoenix AZ	1-800-255-3738
Dynatech Precision Sampling Corp	Baton Rouge LA	1-800-828-1653
Eagle-Picher Environmental Services	Miami OK	1-800-331-7425
Ecological Research and Management	Allen TX	1-800-228-3764
Eddy Current International Inc.	Houston TX	1-800-423-6332
Elite Electronic Engineering	Downers Grove IL	1-800-354-8311
EnCheck		1-800-432-0603
Engine Research	San Lorenzo CA	1-800-445-1479
Engineering & Testing Services Inc.	Indianapolis IN	1-800-229-3871
Engineering Design and Testing	Columbia SC	1-800-338-5227
Enseco-Wadsworth/ALERT Laboratories	North Canton OH	1-800-966-9396
Entropy Environmentalists Inc.	Research Triangle Park NC	1-800-486-3550
Enviro Bio Tech	Bernville PA	1-800-562-2631
EnviroMed Laboratories	Ruston LA	1-800-256-4362
Environ Express Labs	LaPorte TX	1-800-880-0156
Environ Labs	Minneapolis MN	1-800-826-3710
Environmental Conservation Laboratory	Orlando FL	1-800-538-3626
Environmental Health Lab	Macon GA	1-800-841-8919
Environmental Health Lab	South Bend IN	1-800-332-4345
Environmental Laboratories Inc.	Baton Rouge LA	1-800-735-2208
Environmental Quality Lab	Sterling Heights MI	1-800-368-5227
Environmental Reference Lab	Baltimore MD	1-800-368-2576
Environmental Science & Engineering	Peoria IL	1-800-ESE-1999
Environmental Science Lab	Bellingham MA	1-800-343-0707
Environmental Service Lab	Decatur IL	1-800-442-1375
Environmental Science Corp	Middletown CT	1-800-966-9774
Environmental Testing and Consulting Inc	Memphis TN	1-800-528-7688
Environmentrics		1-800-333-3278
Enviroscan Corp	Rothschild WI	1-800-338-7226
EnviroSurv Inc.	Arlington VA	1-800-243-3580
Envisage Environmental Inc.	Richfield OH	1-800-878-0990
The ERM Group	Exton PA	1-800-544-3117
EOHSS	W Lafayette IN	1-800-541-1870
ETL Testing Laboratories Inc.	Cortland NY	1-800-345-3851

Everseal Preservation Labs	Bozeman MT	1-800-262-9269
Excel Technologies Inc.	Enfield CT	1-800-543-9832
Express Lab	Middlesex NY	1-800-843-5227
Fairfax Diagnostics	Fairfax VA	1-800-223-2356
Federal Engineering and Testing	Pompano Beach FL	1-800-848-1919
Food Tek	Morris Plains NJ	1-800-648-8114
Forensic and Scientific Testing	Atlanta GA	1-800-225-1302
Freshcheck Inc.	Minneapolis MN	1-800-882-3005
Furness Controls	Charlotte NC	1-800-858-7237
Galson Corp.	E Syracuse NY	1-800-950-0506
Gasser Associates Arizona	Prescott Valley AZ	1-800-782-4336
General Activation Analysis	San Diego CA	1-800-253-9198
General Activation Analysis	San Diego CA	1-800-367-0526
Genescreen	Dallas TX	1-800-752-2774
Genetic Design Instrumentation	Greensboro NC	1-800-633-9913
Genetics Lab Wound Care	St Paul MN	1-800-328-2634
Gentest	Woburn MA	1-800-334-5229
George F. Nelson	St Joseph MO	1-800-531-9770
Gilson Co. Inc.	Worthington OH	1-800-444-1508
Glitsch Field Services NDE Inc.	N Canton OH	1-800-321-0878
Gould Energy	Cresson PA	1-800-227-1167
Gould Energy	Cresson PA	1-800-634-0879
GT Environmental Lab	Concord CA	1-800-423-7143
GT Environmental Lab	Milford NH	1-800-522-4835
GTEL Environmental Lab	Wichita KS	1-800-633-7936
GTEL Environmental Lab	Milford NH	1-800-441-4835
Gulf South Environmental Laboratory	New Orleans LA	1-800-292-4735
H & H X-Ray Services Inc.	W Monroe LA	1-800-551-5093
Hach Company	Loveland CO	1-800-227-4224
Hampton-Clarke Inc./Veritech	Butler NJ	1-800-426-9992
Halliburton NUS	Pittsburgh PA	1-800-228-6870
Hauser Laboratories Inc.	Boulder CO	1-800-241-2322
Hayden Environmental Group	Miamisburg OH	1-800-548-4031
Health Economic Inc.	Miami FL	1-800-522-7020
Herguth Labs/Oil Research	Vallejo CA	1-800-645-5227
Heritage Environmental Services Inc.	Indianapolis IN	1-800-827-0476
Hernasco Testing Laboratory Inc.	Hudson FL	1-800-642-5270
Histology Consultation Services	Oak Harbour WA	1-800-732-8158
Holmes Laboratory Inc.	Millersburg OH	1-800-344-1101
Human Resource Development USA Inc.	Boca Raton FL	1-800-428-3784
Huntingdon Analytical Services	Middleport NY	1-800-788-5227
Huron Valley Lab Inc.	Romulus MI	1-800-231-4854
H2O Chemists Inc.	Gilbert AZ	1-800-446-3107
H2O Chemists Inc.	Sussex WI	1-800-833-2334
Iatric/Bio Products	Tempe AZ	1-800-662-7846
Iatric Corp	Tempe AZ	1-800-528-4401
ICM Industrial Chemical Measurement	Hillsboro OR	1-800-262-3668
IML		1-800-828-1407
Immgen	College Station TX	1-800-373-5115
Independent Testing Lab	Costa Mesa CA	1-800-962-8378
Independent Testing Lab Inc.	Houston TX	1-800-643-8136
Industrial Compliance	Golden CO	1-800-444-9955
Industrial NDT	Hammond IN	1-800-245-4212
Industrial NDT	Garden City GA	1-800-292-5201
Industrial NDT Co. Inc.	N Charleston SC	1-800-845-1923
Industrial NDT Co. Inc.	Mobile AL	1-800-362-9745
Industrial NDT Inc.	Deer Park TX	1-800-227-3148
Industrial NDT Inc.	Shreveport LA	1-800-435-0663

Industrial NDT Inc.	N Augusta SC	1-800-832-3460
Industrial Performance Testing Inc.	Bristol TN	1-800-228-1478
Inorganic Ventures	Lakewood NJ	1-800-669-6799
Integrated DNA Technologies	Coralville IA	1-800-328-2661
Intellitool Inc.	Batavia IL	1-800-227-3805
Interamerican Management	Cozad NE	1-800-662-2938
International Technology Corp.	Torrance CA	1-800-421-5574
International Testing Labs	Newark NJ	1-800-982-2171
InVitro International	Irvine CA	1-8002INVITRO
JH Consultants	Loudonville NY	1-800-826-9209
J & L Engineering Inc.	Canonburg PA	1-800-453-3608
JACA Corp.	Fort Washington PA	1-800-292-2510
Jackson Immunoresearch Lab	W Grove PA	1-800-367-5296
Johns Hopkins University/ The DACI Lab	Baltimore MD	1-800-344-3224
Johnson Spectra Lab	Mechanicsburg PA	1-800-438-5227
KAG Lab International	Oshkosh WI	1-800-356-6045
Kawin Charles C	Buffalo NY	1-800-245-2946
Kawin Charles C	Broadview IL	1-800-645-2946
Kellco Services	Sacramento CA	1-800-533-5662
Keystone Lab	Newton IA	1-800-858-5227
Keystone Lab Inc.	Asheville NC	1-800-635-5765
Klamath Environmental Services	Klamath Falls OR	1-800-262-5993
Kontes Chemistry and Life Sciences Prod	Vineland NJ	1-800-223-7150
Kontes Glass	Hayward CA	1-800-255-1672
Kronus	Capistrano Beach CA	1-800-822-6999
Kyoto Electronics	Millburn NJ	1-800-458-3168
Laboratory for Diagnostic and Analytical	Indianapolis IN	1-800-253-5322
Laboratory Specialists	Belle Chasse LA	1-800-433-3823
Laboratory Supply	Louisville KY	1-800-626-5275
Laidlaw Environmental Services Inc.	Columbia SC	1-800-356-8570
Larron Laboratory Cape	Girardeau MO	1-800-334-2336
Lawrence Lab	Holbrook NY	1-800-443-9854
Leader Industries	Portage IN	1-800-437-6122
Lee Wan and Associates	Atlanta GA	1-800-433-1091
Lerner Laboratory	Pittsburgh PA	1-800-253-7637
Lis & K Chem Lab	Dorchester MA	1-800-441-6476
Litton Pathology Associates	W Des Moines IA	1-800-258-4429
Lone Star Evaluation Laboratory Inc.	Georgetown TX	1-800-535-5735
Longview Inspection Inc.	Pensacola FL	1-800-258-4773
M R T Lab	Hackensack NJ	1-800-631-1379
MSE-Multitech Services	Butte MT	1-800-441-8213
Magna Chek	Warren MI	1-800-582-8947
Marine & Environmental testing Inc.	Portland OR	1-800-688-5838
Mary Paul Lab	Sparta NJ	1-800-548-1874
Material Testing Corp	Auburn WA	1-800-551-9009
Materials Analytical Services	Norcross GA	1-800-421-8451
Materials Research Lab Inc.	Fords NJ	1-800-424-1776
Matrix Analytical Inc.	Hopkinton MA	1-800-362-8749
MBA Labs	Houston TX	1-800-472-1485
Mc NDT Typeline Ltd	Joliet IL	1-800-253-2970
McKnight Testing	Tampa FL	1-800-223-0968
Med Chek Lab		1-800-548-1723
Medtox Bioanalytical	Chatsworth CA	1-800-331-8670
Medtox Laboratories	New Brighton MN	1-800-832-3244
Memphis Pathology Laboratory	Memphis TN	1-800-423-0504
Memphis Pathology Laboratory	Memphis TN	1-800-423-4491
Meris Laboratories Inc.	Arcadia CA	1-800-782-5663
Metric Medical Lab	Southfield MI	1-800-842-4014

Metuchen Analytical	Edison NJ	1-800-848-4522
Micro Methods	Pascagoula MS	1-800-223-7566
Microanalytical Lab	Gainesville FL	1-800-358-3160
MQS Inspection	Pittsburgh PA	1-800-482-5566
MQS Inspection Inc.	Roseville MN	1-800-622-7616
Munhall Co.	Worthington OH	1-800-247-6629
Municipal Testing Lab	Hicksville NY	1-800-522-6850
National Genetics Institute	N Hollywood CA	1-800-352-7788
National Laboratories Inc.	Evansville IN	1-800-444-4119
National Laboratory Center Inc.	Memphis TN	1-800-526-6339
National Mobile Services Inc.	Floral Park NY	1-800-843-0322
National Testing Laboratories		1-800-458-3330
NDT Distributing	Atlanta GA	1-800-848-0622
Net	Bartlett IL	1-800-654-0473
Neuroscience Associates	Knoxville TN	1-800-972-3401
New Hampshire Materials Lab	Dover NH	1-800-334-5432
NHL National Reference Lab	Nashville TN	1-800-237-7904
Non Destructive Evaluation International	Davidson NC	1-800-892-4873
North American Weather Consultants	Salt Lake City UT	1-800-658-8493
Northeastern Analytical Corp	Marlton NJ	1-800-622-5080
Northern Illinois University Public Opinion	De Kalb IL	1-800-874-1990
Northwest Laboratories Inc.	Berlin CT	1-800-654-1230
Northwest Toxicology Inc.	Salt Lake City UT	1-800-322-3361
OHM Corp.	Findlay OH	1-800-537-9540
Oil Analysis Lab	Spokane WA	1-800-541-1619
Omega Tech Inc.	Trout Dale VA	1-800-437-1404
Oncotech	Irvine CA	1-800-662-6832
Optronic Laboratories Inc.	Orlando FL	1-800-899-3171
PDLA	S Plainfield NJ	1-800-237-7352
PACE Inc.	Golden CO	1-800-878-3434
PACS Laboratory	Industry PA	1-800-367-2587
Pan American Laboratories	Brownsville TX	1-800-382-2487
Partek Laboratories	Houma LA	1-800-445-4619
Particle Diagnostics	San Diego CA	1-800-345-2037
Path Labs	Los Alamitos CA	1-800-762-6062
Perstop Analytical Environmental	Wilsonville OR	1-800-262-3668
Pharmchem Lab Inc.	Menlo Park CA	1-800-446-5177
Pocono Rabbit Farm and Lab Inc.	Canadensis PA	1-800-622-6381
Porous Material	Ithaca NY	1-800-332-1764
Precision Analytical Lab Inc.	Milwaukee WI	1-800-438-9186
Precision Environmental Lab Inc.	Miramar FL	1-800-522-8550
		1-800-331-5408
Princeton Testing Laboratory	Princeton NJ	1-800-548-TEST
Priority One Testing Laboratory	Kearny NJ	1-800-522-0688
Pro Lab Inc.	Round Rock TX	1-800-522-7740
Pro-Tech Oil Analysis Laboratory	Sparks NV	1-800-524-7848
Product Safety Consulting	Bensenville IL	1-800-233-7738
Professional Analytical & Consulting Service	Coraopolis PA	1-800-367-2587
Professional Environmental Services	Davenport IA	1-800-544-5904
Professional Medical Resources	Troy NC	1-800-257-8589
Psychemedics Corp	Boston MA	1-800-522-7424
Psychometrics Systems Inc.	Brookville OH	1-800-541-6026
Pulver Laboratories	Boise ID	1-800-635-3050
Quality Biotech Inc.	Camden NJ	1-800-628-3828
Quantix	Cinnaminson NJ	1-800-322-5487
R & D Laboratory	Columbus OH	1-800-228-4865
Radon Diagnostics Lab	S Portland ME	1-800-992-0150
Recra Environmental	Amherst NY	1-800-52RECRA

Restek	Bellefonte PA	1-800-356-1688
Rollins Environmental Services	Wilmington DE	1-800-992-7837
Roote & Norton Assayers	Silverton CO	1-800-872-3009
Ropak Lab	Irvine CA	1-800-647-6725
Rosemount Analytical Inc.	Santa Clara CA	1-800-538-7708
Ross Analytical Services Inc.	Strongsville OH	1-800-325-7737
Rowlab Science Center	Jacksonville FL	1-800-223-8036
Schwab & Associates	Jacksonville FL	1-800-233-3708
Scientific Resources	N Brunswick NJ	1-800-637-7948
Sensidyne Inc.	Clearwater FL	1-800-451-9444
Servi-Tech Lab	Hastings NE	1-800-468-5411
Sheehan and Co. Inc.	Lakewood CO	1-800-367-1805
Sherry Lab	Muncies IN	1-800-874-3563
Skarshaug Testing Laboratory Inc.	Ames IA	1-800-367-0480
Skelly & Loy Inc.	Harrisburgh PA	1-800-892-6532
Sorrells Research Associates Inc.	Little Rock AR	1-800-331-8139
South Texas Laboratory of Pathology	Westaco TX	1-800-537-5968
Southern Scientific Inc.	Micanopy FL	1-800-633-2408
Southwell Lab	Oklahoma City OK	1-800-872-5669
Spectrex Corp.	Redwood City CA	1-800-822-3940
Spectrum Lab	Ft Lauderdale FL	1-800-262-5983
Stan A. Humber Consultants Inc.	New Lenox IL	1-800-383-0468
Standard Lab	Ashland KY	1-800-842-4886
Standard Lab	S Charleston WV	1-800-624-8668
Standard Lab Inc.	Buckhannon WV	1-800-223-0931
Standards Testing Lab Inc.	Massilon OH	1-800-833-8527
Stay Hard Lab	El Paso TX	1-800-382-0423
Stellar Bio Systems	Columbia MD	1-800-962-6790
Stratagene Co	La Jolla CA	1-800-424-5444
Structure Probe Inc.	West Chester PA	1-800-2424-SPI
Suburban Water Testing Labs		1-800-433-6595
Surface Science Laboratories	Mountain View CA	1-800-321-4775
Surfaces Research and Applications	Lenexa KS	1-800-328-8221
TECO (wood products)	Eugene OR	1-800-628-1763
Tektagen Inc.	Malvern PA	1-800-648-6682
Tenco Environmental Labs	Schererville IN	1-800-643-1835
Testco Inc.	Greensboro NC	1-800-852-6588
Texas Health Department Laboratory	Austin TX	1-800-228-4026
Thermo Analytical Inc.	Monrovia CA	1-800-8777TMA
Thermo Analytical Inc.	Richmond CA	1-800-841-5487
TI Lab	Tulsa OK	1-800-424-9081
TN Technologies Inc.	Round Rock TX	1-800-736-0801
Top Soil Testing Service	Frankfort IL	1-800-458-2530
Toxikon	Woburn MA	1-800-458-4141
Trace America	Miami FL	1-800-872-2313
Trace and Analytics	Austin TX	1-800-247-1024
TRC Environmental	Windsor CT	1-800-TRC-5601
Tri-State Laboratories	Youngstown OH	1-800-523-0347
Triangle Environmental Services	Durham NC	1-800-367-4862
Trinity Environmental Technologies Inc.	Mound Valley KS	1-800-722-8378
Truton US Ltd.	Latham NY	1-800-638-3926
Twin Ports Testing	Superior WI	1-800-373-2562
UIC	Joliet IL	1-800-346-3842
US Test	Lafayette LA	1-800-327-7846
United States Testing Co.	Hoboken NJ	1-800-777-8378
United States Testing Co.	Hoboken NJ	1-800-522-6397
		1-800-346-5926
Universal Laboratory	Garland TX	1-800-447-3996

Valley Toxicology	Sacramento CA	1-800-521-0372
Vanderbilt Reference Laboratory	Nashville TN	1-800-551-5227
Versar Inc.	Springfield VA	1-800-283-7727
Vicam	Somerville MA	1-800-338-4381
Wastex Industries of New Jersey	Bridgeport NJ	1-800-242-0380
Webb Technical Group Inc.	Raleigh NC	1-800-548-7687
West Penn Testing Lab Inc.	Turtle Creek PA	1-800-367-9785
Western Forensic Sciences	Grand Junction CO	1-800-537-8609
Whitbeck Lab	Springdale AR	1-800-874-8195
Willamette Industries Inc.	Charlotte NC	1-800-523-1379
Wingerter Labs	N Miami FL	1-800-345-7645

LABORATORY EQUIPMENT & MONITORING EQUIPMENT

American Sigma	Medina NY	1-800-635-4567
Andersen Instruments Inc.	Atlanta GA	1-800-241-6898
Antek Instruments Inc.	Houston TX	1-800-365-2143
Arts Manufacturing and Supply	American Falls ID	1-800-635-7330
Brainard-Kilman Drilling Company	Stone Mountain GA	1-800-241-9468
CAE Instrument Rental	Palatine IL	1-800-553-5511
Drillers Service Inc.	Hickory NC	1-800-334-2308
Dynatech Precision Sampling Corp	Baton Rouge LA	1-800-828-1653
EnSys Inc.	Research Triangle Park NC	1-800-242-7472
Environmental Instruments	Concord CA	1-800-648-9355
Hach Company	Loveland CO	1-800-227-4224
Hazco Services Inc.	Dayton OH	1-800-332-0435
Inorganic Ventures Inc.	Lakewood NJ	1-800-669-6799
Intek Corp	Houston TX	1-800-323-6527
Milton Roy Company	Rochester NY	1-800-654-9955
Pollulert Systems	Indianapolis IN	1-800-343-2126
SKC - West	Fullerton CA	1-800-752-9378
Tech Line Instruments	Fond Du Lac WI	1-800-328-7518
Tidel Engineering Inc.	Carollton TX	1-800-678-7577

LEAK DETECTION - ENVIRONMENTAL ALARMS

CCI Controls	Southgate CA	1-800-521-5228
Dover Corp	Cincinnati OH	1-800-422-2525
Fugitive Emissions Control	Long Beach CA	1-800-338-3205
Quadrel Services Inc.		1-800-878-5510
Veeder-Root Env Products	Simsbury CT	1-800-873-3313

LINERS & GEOTEXTILES

Atlantic Construction Fabrics	Richmnond VA	1-800-448-3636
Bid D Lining Systems Company	Midland TX	1-800-562-4440
Gundle Lining Systems Inc.	Houston TX	1-800-435-2008
Morton Polymer Systems	Chicago IL	1-800-257-9596
MPC Containment Systems Ltd	Chicago IL	1-800-621-0146
National Seal Co.	Aurora IL	1-800-323-3820
Polymer Fabrication Company	Arab AL	1-800-877-7659

Reef Industries Inc.	Houston TX	1-800-231-2417
Staff Industries Inc.	Farmington Hills MI	1-800-526-1368

LOBBYISTS - REGISTERED ENVIRONMENTAL

Avon Products Inc.	Washington DC	1-800-858-8000
CSX Corporation	Washington DC	1-800-232-0143
General Electric Company	Miami FL	1-800-626-2000

MAPS

ADC The Map People	Bethesda MD	1-800-232-6277
A Metro Zip and Wall Map Co	Peachtree GA	1-800-647-6277
American Map	Maspeth NY	1-800-432-6277
Arrow Publishing Co. Inc.	Paxton MA	1-800-343-7500
Arrow Publishing Co. Inc.	Paxton MA	1-800-367-2627
B & L Distributors	Colorado Springs CO	1-800-648-0565
Carolina Maps	Greenville NC	1-800-248-6277
Cartographics	Monroe CT	1-800-682-4708
Celestial Products	Middleburg VA	1-800-235-3783
Champion Map Rand McNally	Daytona Beach FL	1-800-874-7010
Champion Map Rand McNally	Marlboro MA	1-800-922-9380
The Chart House	Studio City CA	1-800-322-1866
Charts Ink Co	Cottonwood AZ	1-800-423-5094
Datamap Inc.	Eden Prairie MN	1-800-533-7742
Delorme Maps		1-800-227-1656
Earthwalk Press	Eureka CA	1-800-828-6277
ERIIS		1-800-989-0402
First Frame Graphics	Easton MD	1-800-752-4481
First States Map and Globe Co.		1-800-327-7992
Franklin Maps	King of Prussia PA	1-800-356-8676
George F Cram Inc.	Indianapolis IN	1-800-227-4199
Gousha HM	Hendersonville NC	1-800-421-7308
Highway Maps		1-800-255-7608
Intela-Map Inc.	St Joseph MI	1-800-545-4988
International Map Service	Lakewood CO	1-800-426-8676
Keith Map Service Maps Unlimited	Mobile AL	1-800-342-6277
Kingfisher Maps	Seneca SC	1-800-326-0257
King of the Road Map Service Inc.	Mountlake Terrace WA	1-800-223-8852
Map Express	Lakewood CO	1-800-627-0039
Map Merchandisers of Florida	Sarasota FL	1-800-338-5940
The Map Store	Columbus OH	1-800-332-7885
Map Supply Inc.	Lexington NC	1-800-258-9235
Map Supply of Texas	Eastland TX	1-800-457-4627
Mapsource	St Petersburg FL	1-800-262-7123
McCurnin Nautical Maps	Metaire LA	1-800-638-4544
Mc Distributing	Canfield OH	1-800-533-9618
Midland Map	Midland TX	1-800-592-4650
Nationwide Distributor	Miami FL	1-800-327-3108
Nationwide Distributor	Miami FL	1-800-432-1730
Nystrom	Chicago IL	1-800-621-8086
Odyssey Map Store	Indianapolis IN	1-800-972-1388
Pacific Crest Distributors	Redmond WA	1-800-321-8392

Paul R Stoney Print & Map Seller	Williamsburg VA	1-800-732-4923
Pittmon Maps	Portland OR	1-800-547-3576
Pittmon Maps	Portland OR	1-800-452-3228
Raisz Landform Maps	Cambridge MA	1-800-242-3199
Rand McNally Champion Map	Charlotte NC	1-800-438-7406
Raven Maps and Images	Medford OR	1-800-237-0798
Rockwell Enterprises	Carson CA	1-800-331-0293
Southwest Map	Garland TX	1-800-633-1723
Southwinds Press	Tallahassee FL	1-800-824-7784
The Mapp Store (Wholesale)	Columbus OH	1-800-862-7626
United Map Service	Spokane WA	1-800-682-3615
US Geological Survey	Reston VA	1-800USAMAPS

MEDICAL WASTE

BFI Medical Waste System	Dacona CO	1-800-447-8544
BFI Medical Waste Systems of Arizona Inc.	Phoenix AZ	1-800-451-1405
BFI Medical Waste Systems of Utah Inc.	Salt Lake City UT	1-800-321-9993
Browning Ferris Medical Waste Systems	Ft Lauderdale FL	1-800-437-4447
Integrated Medical Waste Management	Hicksville NY	1-800-543-7942
Med Waste Management National	W Springfield MA	1-800-932-9911
Medical Disposal Systems Inc.	Burnsville MN	1-800-235-0137
Medical Waste Specialists Inc.	Brockway PA	1-800-972-6336
Medisin	Prestonburg KY	1-800-822-8012
MEDX	Miami FL	1-800-527-0666
MEDX Inc.	Princeton NJ	1-800-448-4710
Midwest Medical Disposal	Owensville MO	1-800-874-2057
National Medical Waste of Tennessee	Nashville TN	1-800-424-7428
Ortho Medical Waste Management	Daytona Beach FL	1-800-633-0868
P & D Biomedical Waste Disposal	Kingman AZ	1-800-551-6248
SES Medi Waste	Garden Grove CA	1-800-448-0557
Safety First Medical Waste Management Inc.	Bronx NY	1-800-432-3241
Sci-Med Waste Systems	Winston-Salem NC	1-800-662-0088
Tri-State Medical	Hudson NH	1-800-338-0476

NUCLEAR ENERGY - CONSULTANTS, EQUIPMENT & SUPPLY

ABB Power Systems Energy Services		1-800-243-7085
ABB Power Systems Energy Services		1-800-243-8113
Air Systems International		1-800-866-8100
Applied Radiological Control Inc.	Kennesaw GA	1-800-241-6575
Aptech	Sunnyvale CA	1-800-477-2228
Astro Nuclear Dynamics Inc.	Pittsburgh PA	1-800-952-3864
Benthos Robotice Systems Technology	N Falmouth MA	1-800-446-1222
BWNT	Lynchburg VA	1-800-245-1467
CM Research	League City TX	1-800-262-1267
Comtronics Industrial Communications Sy	Lexington KY	1-800-264-6021
Daden Products	Latrobe PA	1-800-537-6007
EG & G Nuclear Instruments Hotline	Oak Ridge TN	1-800-251-9750
F & J Specialty Products	Miami FL	1-800-832-5037
General Atomics	San Diego CA	1-800-854-2233

Griffolyn/Reef Industries	Houston TX	1-800-231-6074
Jeff McManus		1-800-438-5333
National Technical Systems	Acton MA	1-800-723-2NTS
NSSW Numanco	Campbelltown PA	1-800-322-5689
Nuchemco (Industrial Hygiene Consultants)		1-800-682-4362
Nuclear Assurance	Norcross GA	1-800-241-0507
Nuclear Consulting Service Inc.		1-800-992-5192
Nuclear Energy Services	Apex NC	1-800-633-9122
Nuclear Energy Services	Apex NC	1-800-633-9434
Nuclear Energy Services	Apex NC	1-800-848-7795
Nuclear Power Outfitters		1-800-252-8777
Nuclear Support Service	Campbelltown PA	1-800-338-7333
Nuclear Support Service	Munster IN	1-800-322-5689
Nutherm	Mt Vernon IL	1-800-851-7593
Operations Engineering	Fremont CA	1-800-634-3393
Organization for Economic Cooperation	Washington DC	1-800-456-6323
Packard Instrument Company	Meridien CT	1-800-323-1891
Phillips Reliance Mechanical Company	Fort Mill SC	1-800-735-5732
Radiological Inspection Reports	Rockville MD	1-800-854-9052
Radiological Training Services	Burke VA	1-800-222-4716
RPM	Santa Ana Heights CA	1800RADWASTE
Stone and Webster	Boston MA	1-800-421-3042
Unipub	Lanham MD	1-800-274-4888
Unistrut Metal Framing	Wayne MI	1-800-521-7730
Vermont Yankee Nuclear Power Corp	Brattleboro VT	1-800-322-0242

OCCUPATIONAL SAFETY & HEALTH CONSULTANTS

Agio Group	Shawnee Mission KS	1-800-892-3015
AIG Consultants	Parsippany NJ	1-800-972-4420
ALMA Associates		1-800-HELP-572
ATEC Associates	Indianapolis IN	1-800-473-0194
Applied Science & Technology Inc.	Ann Arbor MI	1-800-395-2784
Atlantic Environmental Inc.	Dover NJ	1-800-344-4414
Bear Safety Consulting Inc.		1-800-362-6226
Behavior Science Technology Inc.		1-800-548-5781
Benn Safety Management & Training	Attleboro Falls MA	1-800-282-2366
Bruel & Kjaer Instruments	Decatur GA	1-800-252-4871
Canadian Center for Occupational Health and Safety ON CN		1-800-668-4284
Concord Geotechnical	Wheatridge CO	1-800-635-9582
Consortium for EMP Impairment Control	Oxford OH	1-800-354-0925
Corporate Safety & Health Services Inc.		1-800-533-CSHS
Crawford Risk Control Services		1-800-723-3890
CURA Inc.	Dallas TX	1-800-486-7117
Diamond Safety Concepts	Enon OH	1-800-437-0779
DuPont Safety & Environmental Resources	Wilmington DE	1-800-532-7233
Earthquake Preparedness Society	Downey CA	1-800-628-9111
Ebasco Environmental	Lyndhurst NJ	1-800-580-3765
ELB & Associates	Chapel Hill NC	1-800-334-5478
EMCON Associates	San Jose CA	1-800-753-6266
Engineering Safety Consultants	San Antonio TX	1-800-327-1516
Ennis Lumsden Boylston & Associates	Cowpens SC	1-800-442-0193
Environmental Health & Engineering Inc.	Newton MA	1-800-825-5343
The Ferguson Group	Seattle WA	1-800-722-2004
Florida Employers Safety Association	Lakeland FL	1-800-433-3971

Galson Corp.	E Syracuse NY	1-800-950-0506
Global Loss Control Associates Inc.	Easton PA	1-800-962-7850
GNA	Grand Haven MI	1-800-968-1611
Graseby Anderson	Atlanta GA	1-800-241-6898
HNU Systems Inc.	Newton MA	1-800-724-5600
Harding Lawson Associates	Novato CA	1-800-578-0821
Hazco Services Inc.	Dayton OH	1-800-332-0435
Hazcon Inc.	Missoula MT	1-800-528-7657
Health Advancement Services	Tempe AZ	1-800-782-5500
Hurd Ben & Associates	Arcata CA	1-800-472-1757
ILC Dover Inc.	Fredericka DE	1-800-631-9567
Industrial Health Inc.		1-800-221-1330
Industrial Safety Services N.C.		1-800-348-7534
Industrial Scientific Corp.	Oakdale PA	1-800-338-3287
Intek Corp.	Houston TX	1-800-323-6527
Interstate Safety Engineering	Orange CA	1-800-221-8568
I.S.H. Inc.		1-800-488-6399
KBN Engineering & Applied Sciences Inc.	Gainesville FL	1-800-333-4526
L & L Transportation Services	Appleton WI	1-800-852-1179
Mansdorf & Associates Inc.	Stow OH	1-800-331-3044
Medical Resource Services	Philadelphia PA	1-800-547-2045
Medinomics Inc.	Pinehurst NC	1-800-392-4598
Michigan Fire Safety Foundation	Michigan Center MI	1-800-242-4638
Miller Equipment		1-800-873-5242
Mobile Health Services	Cheswick PA	1-800-MED1197
Most Health Services Inc.	Voorhees NJ	1-800-828-6678
National Medical Advisory Service	Bethesda MD	1-800-258-0014
National Safety Council	Itasca IL	1-800-621-7619
Network Environmental Systems Inc.	Folsom CA	1-800-637-2384
NJ & Associates Inc.	Dallas TX	1-800-622-0177
Outokumpu Electronics Inc.	Langhorne PA	1-800-229-9209
Professional Safety Royal	Palm Beach FL	1-800-562-7233
Professional Service Industries Inc.	Lawrence KS	1-800-548-9166
Roche Analytics Laboratory	Richmond VA	1-800-888-8061
Safety Line Consultants	Chesterfield MO	1-800-952-1363
Safety Management	Norfolk VA	1-800-932-8053
Safety Management Corp.	S Burlington VT	1-800-639-1920
Safety Training Specialist	Noblesville IN	1-800-435-8850
Schaible Associates	Mount Joy PA	1-800-832-5564
Scott Lawson Group Ltd.	Concord NH	1-800-645-7674
Siemens Hearing Conservation Group		1-800-865-3277
Spectrex Corp.	Redwood City CA	1-800-822-3940
Spectrum Safety Products	Gardena CA	1-800-772-8786
STC Enterprises	Muscatine IA	1-800-382-1361
Sverdup Environmental Inc.	Maryland Heights MO	1-800-325-7910
Texas Safety Associates		1-800-723-3895
3M Occupational Health & Environmental	St Paul MN	1-800-328-1667
Traffic Safety Consultants Inc.	Richmond VA	1-800-628-9547
Western Pennsylvania Safety Council	Cherry Hill NJ	1-800-245-9772
Workers Compensation Solutions Inc.	Ft Wayne IN	1-800-572-3359

ODOR CONTROL

Air Scent International	Braddock PA	1-800-247-0770
Microban Germicide Company	Braddock PA	1-800-556-0111

PESTICIDES, HERBICIDES, AND INSECTICIDES - INFORMATION

Agent Orange Veteran Payment Program	Hartford CT	1-800-225-4712
	Birth Defect Information Line	1-800-922-9234
National Pesticides Information Clearinghse	Lubbock TX	1-800-858-7378
Vietnam Veterans Agent Orange Victims	Darien CT	1-800-521-0198

POISON CONTROL CENTERS

ALABAMA

Alabama Poison Center	Tuscaloosa AL	1-800-462-0800
Regional Poison Center	Birmingham AL	1-800-292-6678

ALASKA

Anchorage Poison Center	Anchorage AK	1-800-478-3193

ARIZONA

Arizona Poison and Drug Information Center	Tucson AZ	1-800-362-0101

ARKANSAS

Arkansas Poison and Drug Information Center	Little Rock AR	1-800-482-8948

CALIFORNIA

Fresno Regional Poison Control Center	Fresno CA	1-800-346-5922
Los Angeles County Regional Poison Contr	Los Angeles CA	1-800-777-6476
San Diego Regional Poison Control Center	San Diego CA	1-800-876-4766
San Francisco Bay Area Regional Poison	San Francisco CA	1-800-523-2222
Santa Clara Valley Regional Poison Control	San Jose CA	1-800-662-9886
Univ of CA Davis Regional Poison Control	Sacramento CA	1-800-342-9293
Univ of CA Irvine Regional Poison Control	Orange CA	1-800-533-4404

COLORADO

Rocky Mountain Poison and Drug Center	Denver CO	1-800-332-3070
	Montana	1-800-525-5042
	Nevada	1-800-446-6179

CONNECTICUT

Connecticut Poison Control Center	Farmington CT	1-800-343-2722

FLORIDA

Florida Poison Information Center	Tampa FL	1-800-282-3171

GEORGIA

Georgia Regional Poison Control Center	Atlanta GA	1-800-282-5846

HAWAII
Hawaii Poison Center   Honolulu HI   1-800-362-3585

IDAHO
Idaho Emergency Medical and Poison Control
                                                        Boise ID   1-800-632-8000

ILLINOIS
Central and Southern Illinois Regional   Springfield IL   1-800-252-2022
Chicago and Northeastern Illinois Region   Chicago IL   1-800-942-5969

INDIANA
Indiana Poison Control Center   Indianapolis IN   1-800-382-9097

IOWA
Poison Control Center   Iowa City IA   1-800-272-6477
Variety Club Poison and Drug Information   Des Moines IA   1-800-362-2327

KANSAS
Mid-America Poison Center   Kansas City KS   1-800-332-6633

KENTUCKY
Kentucky Regional Poison Center   Louisville KY   1-800-722-5725

MAINE
Maine Poison Control Center   Portland ME   1-800-442-6305

MARYLAND
Maryland Poison Center   Baltimore MD   1-800-492-2414

MASSACHUSETTS
Massachusetts Poison Control System   Boston MA   1-800-682-9211

MICHIGAN
Blodgett Regional Poison Center   Grand Rapids MI   1-800-632-2727
                                                        Hearing Impaired   1-800-356-3232
Bronson Poison Center   Kalamazoo MI   1-800-442-4112

MINNESOTA
Minnesota Regional Poison Center   St Paul MN   1-800-222-1222

MISSOURI
Regional Poison Center   St Louis MO   1-800-366-8888

MONTANA
Rocky Mountain Poison and Drug Center   Denver CO   1-800-525-5042

NEBRASKA
The Poison Center — Omaha NE — 1-800-955-9119

NEW HAMPSHIRE
New Hampshire Poison Information Center — Lebanon NH — 1-800-562-8236

NEW JERSEY
New Jersey Poison Information and Education
Newark NJ — 1-800-962-1253

NEW MEXICO
New Mexico Poison and Drug Information — Albuquerque NM — 1-800-432-6866

New York
Central New York Poison Control Center	Syracuse NY	1-800-252-5655
Hudson Valley Regional Poison Center	Nyack NY	1-800-336-6997
Life Line Finger Lakes Regional Poison	Rochester NY	1-800-333-0542
Western New York Poison Control Center	Buffalo NY	1-800-888-7655

NORTH CAROLINA
Duke University Poison Control Center	Durham NC	1-800-672-1697
Triad Poison Center	Greensboro NC	1-800-722-2222
Western North Carolina Poison Center	Asheville NC	1-800-542-4225

NORTH DAKOTA
North Dakota Poison Information Center — Fargo ND — 1-800-732-2200

OHIO
Akron Regional Poison Control Center	Akron OH	1-800-362-9922
Central Ohio Poison Control Center	Columbus OH	1-800-682-7625
Mahoning Valley Poison Center	Youngstown OH	1-800-426-2348
Regional Poison Control System	Cincinnati OH	1-800-872-5111
Western Ohio Poison and Drug Informatio	Dayton OH	1-800-762-0727

OKLAHOMA
Oklahoma Poison Control Center — Oklahoma City OK — 1-800-522-4611

OREGON
Oregon Poison Center — Portland OR — 1-800-452-7165

PENNSYLVANIA
Cental Pennsylvania Poison Center	Hershey PA	1-800-521-6110
Northwest Regional Poison Center	Erie PA	1-800-822-3232
Susquehanna Poison Center	Danville PA	1-800-352-7001

SOUTH CAROLINA
Palmetto Poison Center — Columbia SC — 1-800-922-1116

SOUTH DAKOTA

McKennon Poison Control Center	Sioux Falls SD	1-800-952-0123
McKennon Poison Control Center		1-800-843-0505
St. Luke's Midland Poison Control Center	Aberdeen SD	1-800-592-1889

TENNESSEE

Middle Tennessee Regional Poison/Clinic	Nashville TN	1-800-288-9999

TEXAS

North Texas Poison Center	Dallas TX	1-800-441-0040

UTAH

Intermountain Regional Poison Control	Salt Lake City UT	1-800-456-7707

VIRGINIA

University of Virginia Blue Ridge Poison	Charlottesville VA	1-800-451-1428
Virginia Poison Center	Richmond VA	1-800-552-6337

WASHINGTON

Central Washington Poison Center	Yakima WA	1-800-572-9176
Mary Bridge Poison Center	Tacoma WA	1-800-542-6319
Seattle Poison Center	Seattle WA	1-800-732-6985
Spokane Poison Center	Spokane WA	1-800-572-5842

WEST VIRGINIA

West Virginia Poison Center	Charleston WV	1-800-642-3625

WYOMING

The Poison Center	Omaha NE	1-800-955-9119
St. Luke's Midland Poison Control Center	Aberdeen SD	1-800-592-1889

POLLUTION CONTROL & MEASUREMENT

American Sigma	Medina NY	1-800-635-1230
Andersen Samplers Inc	Smyrna GA	1-800-241-6898
Anderson 2000	Peachtree City GA	1-800-241-5424
B H A Group Inc	Kansas City MO	1-800-821-2222
Baghouse Services Inc	Aiken SC	1-800-437-1168
Bio Systems	Beloit WI	1-800-232-2847
Clean Air Engineering	Palatine IL	1-800-553-5511
Clear 2 O	Northbrook IL	1-800-253-2720
Enercon Systems	Atlanta GA	1-800-572-1835
Enterra Instrumentation Technologies	Exton PA	1-800-634-4046
Fuel Saver Products	Grand Junction CO	1-800-281-5186
Gravity Flow Systems	Carbondale PA	1-800-237-7500
Hoyt Corp	Westport MA	1-800-343-9411
Marine Pollution Control	Detroit MI	1-800-521-8232
Midwesco Filter Resources	Winchester VA	1-800-336-7300
N I Industries	Los Angeles CA	1-800-266-7747
N-Con Systems	Larchmont NY	1-800-932-6266

National Filter Media	Moorestown NJ	1-800-628-5110
O R S Environmental Equip	Greenville NH	1-800-228-2310
Oil Pollution Advisory Services	E Rutherford NJ	1-800-438-6727
Pensacola Pollution Control Inc	Pensacola FL	1-800-642-7445
Precip Tech Inc - A BHA Group Company	Kansas City MO	1-800-336-2585
Seapill Intl	Southold NY	1-800-732-7745
Sengineering	Novato CA	1-800-822-3459
Southern Environmental	Pensacola FL	1-800-282-8734
Staclean Diffuser	Salisbury NC	1-800-222-9119
Staclean Diffuser	Salisbury NC	1-800-438-3850
Telpac Tower Packing	Boston MA	1-800-523-0948
Wahlco	Leawood KS	1-800-228-5910
Zimpro Passavant	Avon Lake OH	1-800-421-3144
Zimpro Passavant	Irondale AL	1-800-633-9501

QUALITY MANAGEMENT

Advanced Performance Solutions Inc.	Arlington VA	1-800-235-1999
Advanced Practical Thinking Training In	Des Moines IA	1-800-621-3366
Affiliated Quality Enterprises Inc.	Lincolnshire IL	1-800-377-4660
American Productivity & Quality Center	Houston TX	1-800-776-9676
American Quality Systems	New Baltimore MI	1-800-776-3090
American Society For Quality Control	Milwaukee WI	1-800-248-1946
American Society For Quality Control (ASQC) Press		1-800-248-1946
Apian Software	Menlo Park CA	1-800-237-4565
Association for Quality and Participation		1-800-733-3310
Aurora Pictures Inc.	Minneapolis MN	1-800-346-9487
Avatar International Inc.	Norcross GA	1-800-282-8274
BQS Inc.	Montvale NJ	1-800-624-5892
Basler/Maltz Associates	Westport CT	1-800-369-6920
The Batten Group		1-800-234-3176
Belgard-Fisher-Rayner Inc.	Beaverton OR	1-800-426-1060
Benchmark Courses in Effective Communic	Altamont NY	1-800-435-3927
Blackhawk Technology		1-800-528-9333
Blanchard Training & Development Inc.	Escondido CA	1-800-728-6000
Blue Ridge Institute	Roanoke VA	1-800-258-3424
David W. Buker	Antioch IL	1-800-654-7990
Bureau Veritas Quality International	Jamestown NY	1-800-937-9311
The Carman Group	Richardson TX	1-800-942-6880
CEEM Information Services		1-800-745-5565
Charthouse International Learning Corp		1-800-328-3789
The Clark Wilson Group	Silver Spring MD	1-800-537-7249
CyberMetrics Corp	Rochester Hills MI	1-800-777-7020
Design Of Experiments (DOE/Direct)	Cambridge MA	1-800-395-6392
Development Dimensions International	Bridgeville PA	1-800-933-4463
Digital Equipment Corp.	Maynard MA	1-800-DECINFO
DuPont Quality Management & Technology	Newark DE	1-800-441-8040
Eastern Michigan University Center for	Ypsilanti MI	1-800-932-8689
Effective Schools Products Ltd.	Okemos MI	1-800-827-8041
Entela Inc. Quality Services Registration	Grand Rapids MI	1-800-888-3787
Executive Adventure	Atlanta GA	1-800-949-8326
Executive Learning Inc.	Brentwood TN	1-800-929-7890
Faxon Research Services		1-800-945-4377
The Forum Corp.	Boston MA	1-800FORUM11
General Physics Corp	Columbia MD	1-800-727-6677

George Washington University	Washington D.C.	
	National Satellite Network	1-800-932-2337
	National Satellite Network - Canada	
		1-800-535-4567
	Continuing Engineering Education Program	
		1-800-424-9773
Georgia Tech Center for International	S Atlanta GA	1-800-859-0968
Goal/QPC		1-800-643-4316
GPS Technologies Inc.	Troy MI	1-800-346-9533
International Quality & Productivity Center	Montclair NJ	1-800-882-8684
International Quality Verification Service	New York NY	1-800-362-3639
The ISO Auditing Network	Altadena CA	1-800-ISO-9999
Juran Institute		1-800-847-7154
Johnson Controls Institute	Milwaukee WI	1-800-524-8540
Joiner Associates Inc.	Madison WI	1-800-669-8326
Jostens Learning		1-800-247-1380
Launsby Consulting	Colorado Springs CO	1-800-788-4363
Managemewnt Institute School of Busines	Madison WI	1-800-292-8964
Manugistics/Statgraphics	Rockville MD	1-800-592-0050
Microsteps	San Diego CA	1-800-JIT-1022
Miles River Press	Alexandria VA	1-800-767-1501
NCS Assessments	Minneapolis MN	1-800-627-7271
National Safety Council	Itasca IL	1-800-621-7619
Neville Clarke Quality		1-800-979-9001
OD & D Seminars	Houston TX	1-800-648-1480
Perry Johnson Inc.	Southfield MI	1-800-800-0450
Productivity-Quality Systems Inc.	Miamisburg OH	1-800-777-3020
Proudfoot Crosby Inc.	Winter Park FL	1-800-245-5751
QCI International	Red Bluff CA	1-800-527-6970
Q-Cee's Products Division/The Mountain	Houston TX	1-800-950-4922
Qualitran Professional Services Inc.		1-800-461-9902
Quality Alert Institute	New York NY	1-800-221-2114
Quality America Inc.	Tucson AZ	1-800-722-6154
Quality Focus Inc.		1-800-992-9079
Quality Inspection Services Co.	Columbia MD	1-800-572-0553
Quality Leaders	Nitro WV	1-800-598-6792
Quality Magazine	Carol Stream IL	1-800-323-5155
Quality Management Assistance Group	Appleton WI	1-800-236-7802
Quality Plus Engineering	Portland OR	1-800COMPETE
Quality Publishing Inc.	Tucson AZ	1-800-628-0432
Quality Resources	White Plains NY	1-800-247-8519
Qualtec Institute for Competitive Advan	N Palm Beach FL	1-800-247-9871
Rich Wilkins & Company	Shepardsville KY	1-800-944-7269
Salenger Inc.	Santa Monica CA	1-800-775-5025
SGS International Certification Service	Hoboken NJ	1-800-777-8378
Shilay Associates Inc.	Wantaugh NY	1-800-541-1366
Skill Dynamics		1800IBMTEACH
Society for Manufacturing Engineers	Dearborn MI	1-800-733-4763
SPC Press		1-800-545-8602
Stanton & Hucko	Rochester NY	1-800-325-4239
Stat-Ease Inc.	Minneapolis MN	1-800-325-9829
Stat Guide	Pittsburgh PA	1-800-949-7942
Stephen Computer Services Inc.	Walled Lake MI	1-800-553-4772
Technicomp		1-800-735-4440
Teleometrics International	The Woodlands TX	1-800-527-0406
Tennessee Associates International		1-800-426-4121
The Training Edge	Palatine IL	1-800-292-4375
The Victoria Group	Fairfax VA	1-800-845-0567

TQM Resources Inc.	Florham Park NJ	1-800-551-4008
TQN Publishing		1-800-836-0325
United Training Media	Niles IL	1-800-424-0364
University of Iowa Publications Order Dept	Iowa City IA	1-800-235-2665
Video Arts Training		1-800-553-0091
Video Training Resource (VTR)		1-800-828-8190
Wermes	Hoffman Estates IL	1-800-532-7687
Zontec Inc.	Cincinnati OH	1-800-955-0088

RADIOACTIVE MATERIALS

Allied Technologies Group	Richland WA	1-800-321-2844
Allied Technology Great Lakes	Lawrence MI	1-800-255-2845
Amersham Corp	Arlington Heights	1-800-323-6695
Radiation Safety & Nuclear Products	Salt Lake City	1-800-537-7767
Ross Ellis	Northridge CA	1-800-447-2149
Test ER	Addison IL	1-800-622-3235

RADIOACTIVE WASTE - MANAGEMENT & DISPOSAL

Chem-Nuclear Systems Inc.	Raleigh NC	1-800-628-6327
Decon International	Bethel Park PA	1-800-332-6648
NDL Org Inc.	Peekskill NY	1-800-635-6351
US Ecology	Louisville KY	1-800-626-5317

RADIOACTIVITY - EQUIPMENT, SUPPLIES & INSTRUMENTATION

Berthold/Ortec	Oak Ridge TN	1-800-251-9750
Dosimeter Corp	Cincinnati OH	1-800-322-8258
EG & G Nuclear Instruments	Oak Ridge TN	1-800-251-9750
EG & G Ortec	Oak Ridge TN	1-800-251-9750
Meter Conversions Inc.	Houston TX	1-800-523-0515
Ohmart Corp	Cincinnati OH	1-800-543-8668
Packard Instrument Company	Meriden CT	1-800-323-1891

RADON

A B E Radiation Measurements Lab	Lenhartsville PA	1-800-336-4153
Aarden Testing and Evaluation Inc.	Sarasota FL	1-800-441-2104
ABE Measurements Lab	'Lenhartsville PA	1-800-336-4153
Alpha Control	Rockville MD	1-800-466-1987
Alpha Spectra Inc	Grand Junction CO	1-800-231-2545
American Improvement	Altoona PA	1-800-922-5857
American Radon Services Ltd	Ames IA	1-800-272-3668
Approved De-Watering & Radon	Cornwall NY	1-800-439-7236
Carolina Home Consultants	Newton NC	1-800-328-4355
Certus Lab	Atlanta GA	1-800-325-5655
Consolidated Radon Labs	Dayton NJ	1-800-872-7236
Ecodex	Lexington MA	1-800-432-6339
Empac	New Haven CT	1-800-422-3662

Evanshire Co. Ltd.	Rome NY	1-800-962-5961
General Health Physics	Lorton VA	1-800-247-6572
JCN Radon Testing and Mitigation	New Columbia PA	1-800-526-2334
Key Technology Inc	Lebanon PA	1-800-523-4964
Mincey Pest Control	Whiteville NC	1-800-542-0989
Monitor Technology Ltd and Enviralert Inc.	San Francisco CA	1-800-354-1149
National Radon & Consulting Group	Glastonbury CT	1-800-334-6724
Niton Inc.	Bedford MA	1-800-875-1578
Northern Illinois Radon Detection	Wheaton IL	1-800-542-0329
Penn-Rad Inc	Elizabeth PA	1-800-232-3995
Prohand Services	Belgrade MT	1-800-847-2366
Property Services	Clarksburg WV	1-800-231-8378
Rad Elec	Richmond VA	1-800-526-5482
Radiation Data	Princeton NJ	1-800-325-3579
Radiation Safety & Control Service	Stratham NH	1-800-525-8339
Radon Central Associates	Bound Brook NJ	1-800-368-6646
Radon Control Systems	Stamford CT	1-800-343-8304
Radon Detection Systems	Boulder CO	1-800-627-2366
Radon Engineering	Mahwah NJ	1-800-327-2366
Radon Mitigation	Irvington NY	1-800-525-3953
Radon Professional Services Inc	Jacksonville Beach FL	1-800-525-7236
Radon Research Group Inc	Burtonsville MD	1-800-445-8436
Radon Testing Corp of America	Irvington NY	1-800-457-2366
Radon X	Keene NH	1-800-572-3669
Radon-One	Columbus OH	1-800-441-5277
Southern Radon	Marietta GA	1-800-537-2366
Wanner Group	Strasburg PA	1-800-326-3091
Women's Center For Radiology	Orlando FL	1-800-367-1870
Worldwide Radon Testing	Pompano Beach FL	1-800-872-2477

RECYCLING CENTERS

A-Tec Recycling Inc	Des Moines IA	1-800-551-4912
Action Disposal	Manor TX	1-800-873-4810
Aluminum Recovery Corp	Yarmouth MA	1-800-227-4226
American Energy Recyclers	Montgomery AL	1-800-264-8503
Atlantic Coast Recycling	Ft Pierce FL	1-800-329-2826
Baltimore Foam Recycle Center	Baltimore MD	1-800-787-3626
C E G Industries	Borger TX	1-800-646-6406
Fiber Mulch Of Georgia	Valdosta GA	1-800-862-4373
Lubrx Products	Stoughton MA	1-800-982-0153
Recovery 1	Tacoma WA	1-800-949-5852
Refrigerant Reclaim Inc	Woodbridge VA	1-800-238-5902
Sajaw Inc	Samson AL	1-800-637-2529
Schaffer Industries	Shenandoah PA	1-800-544-2006
Texas Tire Disposal	Ft Worth TX	1-800-654-8868
Triple S Recycling	Tionesta PA	1-800-859-3693
Waste Recovery	Baytown TX	1-800-249-5087

RECYCLING - PRODUCTS, EQUIPMENT, SERVICES, AND CONSULTING

A & B Recycling And Scrap	Rossville GA	1-800-360-4658
A & M Enterprises-Reel Recycling	Humboldt TN	1-800-643-2425

A1 Recycling Circus	Gilroy CA	1-800-252-0288
Academy Corp	Albuquerque NM	1-800-545-6685
Adams Brown Clermont Recycling	Georgetown OH	1-800-722-7399
Advance Environmental Recycling Corp	Allentown PA	1-800-554-2372
AGI/An-Gun Inc	West Bend WI	1-800-842-8646
Alvin Andrews	New Market VA	1-800-238-2768
Amber Enterprises Toner Cartridge Recycling	Buxton ND	1-800-526-7056
American Diversified Silver Inc.	Anaheim CA	1-800-243-2580
American Fluid Technology	Westford MA	1-800-245-6869
American Hospital Association		1-800-242-2626
American Paper Recycling Corp	Mansfield MA	1-800-422-3220
American Plastics Council Hotline		1-800-2-HELP90
American Recycling Market Directory		1-800-267-0707
American Resource Recovery Ltd	Maywood IL	1-800-841-6900
American Services Corp	Golden Valley MN	1-800-972-7298
Approved Oil Service	Commerce City CO	1-800-525-0850
Ashley Information Group Recycled Toner	S San Francisco CA	1-800-237-5001
Asset Recovery Corp	St Paul MN	1-800-472-2081
Automotive Industrial Recyclers	Longwood FL	1-800-322-2472
Back To Earth Recycling	N Liberty IA	1-800-626-8733
Bastrop Recycling	Bastrop TX	1-800-321-6385
Battaglia Recycling Services	Niagara Falls NY	1-800-822-0035
Black Gold Corp	Nashville TN	1-800-351-0643
Blue Maxx	Detroit Lakes MN	1-800-551-2583
Boro Recycling Inc.	Maspeth NY	1-800-531-2247
Brandywine Recyclers	Lebanon PA	1-800-843-2200
Brewer Cote of Florida Inc.	Miami FL	1-800-752-7603
Brock Steel	Cumberland MD	1-800-348-9568
Business Ecology Products	Walnut Creek CA	1-800-932-2235
C P Mfg	National City CA	1-800-462-5311
C R T Oil Filters Recycling	Bridgewater MA	1-800-833-8278
C.M. Companies	Bridgeview IL	1-800-235-8351
Can-Upper Inc.	Birmingham AL	1-800-436-8476
Casella Waste Management Inc.	Rutland VT	1-800-227-3552
Casella Waste Management Inc.	Rutland VT	1-800-872-3275
Central Florida Auto Crushers	Lakeland FL	1-800-231-6407
Central Iron and Metals	Hamilton IL	1-800-531-9085
Central States Fiber	Shelbyville IN	1-800-222-5049
Central Trading and Recycling	New Richland MN	1-800-445-0961
Circle T Tire Chopping & Recycling	Pearlington MS	1-800-432-8804
Columbus McKinnon	Sarasota FL	1-800-848-1071
Commercial Metals	Chatanooga TN	1-800-243-2181
Commodity Recovery Systems Inc	Thomasville NC	1-800-845-8037
Computerized Waste Systems	Louisville KY	1-800-233-9923
Concord Trading	Atlanta GA	1-800-352-6053
Container Recovery	Des Moines IA	1-800-372-6072
Container Recovery Corp	Nashua NH	1-800-258-1080
Cow Poke Inc.	Sheldon IA	1-800-257-5924
CR Container Services	Walpole MA	1-800-437-2267
CSL & Associates	Dunwoody GA	1-800-622-6069
D & W Sales	Knoxville TN	1-800-225-5439
Delaware Recyclable Products	New Castle DE	1-800-344-3774
Doe Run	Boss MO	1-800-635-5256
Don Zwiers & Associates	Joliet IL	1-800-835-5160
Dyna-Pak Corp	Lawrenceburg TN	1-800-759-3962
Dynex Environmental Inc	St Paul MN	1-800-659-2804
E T I Technologies Inc	Ft Wayne IN	1-800-854-8033
Eagle Recycled Products	Anaheim CA	1-800-448-4409

Ecoservice-Antifreeze	Thomasville GA	1-800-932-8919
Empire Metal Recycling	Huntington WV	1-800-472-9583
Enviresource Management Group	Silver Spring MD	1-800-862-4411
Environmental Management Products	Ridgefield NJ	1-800-422-5551
Environmental Products Ltd.	Ft Worth TX	1-800-722-7593
Eticam	Providence RI	1-800-782-3552
Fauste Oil Service	Irvine KY	1-800-553-5840
Fecon Inc.	Fairfield OH	1-800-528-3113
Fibres International Inc.	Bellevue WA	1-800-356-6793
Fibrex	Chesapeake VA	1-800-346-4458
Filter Recyclers Corp	Arlington WA	1-800-352-4002
Florida Metal Recycling	Delray Beach FL	1-800-548-3896
Forest Ridge Products	Toledo OH	1-800-442-7874
Forest Ridge Products	Worthington OH	1-800-442-7874
Franklin Recycling	New Bedford MA	1-800-645-3587
G & G Auto Recycling	Winchester VA	1-800-368-0854
G & H Mfg Inc	Arlington TX	1-800-654-5291
GDS Recycling Service	Conover NC	1-800-833-7312
Geneva Recycling Separators		1-800-880-1017
Glass Recycling	Woodstock GA	1-800-472-1196
Golden Systems Inc.	Jacksonville FL	1-800-292-6911
Green Earth Recovery Systems	Bloomfield Hills MI	1-800-528-1998
Halon Recycling Corp	Washington DC	1-800-258-1283
Hazleton Scrap Recycling Center	Hazleton PA	1-800-231-3659
Hi Tech Ind Inc.	Northridge CA	1-800-553-0509
Hi-Rise Recycling Systems	Miami FL	1-800-231-3888
Horsehead resources Development	Palmerton PA	1-800-253-5579
Industrial Service Corp-Corporate Hdqrs	Blue Summit MO	1-800-821-4302
Industrial Services America	Louisville KY	1-800-824-2144
Ingolds Hico Inc.	Bel Air MD	1-800-874-6465
Inky Dew	San Diego CA	1-800-262-4659
International Compactor Inc.	Hilton Head Island SC	1-800-423-4003
International Recycling Services Inc.	Grand Island NY	1-800-532-4709
J R Engineering Corp	W Swanzey NH	1-800-732-0123
J S M Recycling	Marble NC	1-800-845-5503
Jeter James	Tyler TX	1-800-742-0090
JR's Appliance Disposal Inc.	Inver Grove Heights MN	1-800-358-6563
Lake Iron and Metal	Hammond IN	1-800-858-5065
Land Reclamation	Westbrook ME	1-800-831-5831
Larry's Radiator Service	Charlotte NC	1-800-730-7340
Laser Technology Toner to Donor	W Chester PA	1-800-362-8663
Laser Vision Inc.	Wilmington NC	1-800-325-4573
Lehigh Valley Recycling	Coplay PA	1-800-331-6937
Lighting Resources Inc.	Pomona CA	1-800-572-9253
Link-A-Bag Systems Inc.	E Syracuse NY	1-800-321-8154
Lipp Brothers	Rapid City SD	1-800-628-5477
LUBRX	North Attleboro MA	1-800-343-3381
Lundell Manufacturing	Cherokee IA	1-800-831-4841
M R Enterprises Recycling Technologies	Ottawa IL	1-800-872-7590
Mainmetal Recycling	Auburn ME	1-800-492-0852
Marine Shale Processors	Amelia LA	1-800-633-0090
Marine Shale Processors	Amelia LA	1-800-872-6774
May Manufacturing Distributing Corp	Denver CO	1-800-292-7968
Michigan Silver Exchange	Ferndale MI	1-800-442-8190
Midwest Power & Equipment	W Chicago IL	1-800-635-2023
Miller Auto Crushing	Sante Fe TX	1-800-982-2385
Mindis Bondi Metals	Baltimore MD	1-800-458-0042
Mobley Co Inc	Kilgore TX	1-800-999-8628

Monarch Specialty Systems Inc	Ossian IN	1-800-537-6260
Morbark Parts & Service Corp Fax Only	Winn MI	1-800-832-5618
Multi Container Recycler	Grand Rapids MI	1-800-898-5558
Nation Wide Recyclers	Polkton NC	1-800-554-1680
National Techno	Collegeville PA	1-800-858-2824
Nebraska State Recycling Association	Grand Island NE	1-800-248-7328
Nedland Industries	Ridgeland WI	1-800-447-4925
NEED	Johnston RI	1-800-447-3873
Neroc	E New Haven CT	1-800-542-2569
New England Crinc	N Billerica MA	1-800-872-7462
New England Crinc	Chicopee MA	1-800-542-2402
Newell Enterprises	San Antonio TX	1-800-883-7448
North East Chemical	Cleveland OH	1-800-843-6322
Nova Sylva Inc.	Sherbrooke QC CN	1-800-561-2963
NWORKS		1-800-547-4226
Omnisphere Corp.	W Springfield MA	1-800-654-6664
Osborn Engineering	Tulsa OK	1-800-672-6761
Pacific Recycling	Billings MT	1-800-932-2267
Peck Recycling	Richmond VA	1-800-762-6484
Peck Recycling Company	Woodford VA	1-800-647-7325
PGM Technologies	Swartswood NJ	1-800-852-0041
Planet Earth Recycling	Philadelphia PA	1-800-327-8450
Priefert Mfg Co Inc	Mt Pleasant TX	1-800-742-0090
Pro-Tainer Inc.	Alexandria MN	1-800-248-7761
R A Shields & Co	Richmond IN	1-800-752-2144
Recon Oil Inc	Odessa TX	1-800-382-9576
Recycle Systems of America Inc.	Greenland NH	1-800-833-3781
Recycled Products Management Inc.	Sanford NC	1-800-223-9021
Recycling Center The	Georgetown OH	1-800-722-7399
Recycling Equip Service	Baxley GA	1-800-423-3885
Recycling Equipment Connection	Garner NC	1-800-752-6124
Recycling Equipment Connection	FAX Line	1-800-232-9732
Recycling Equipment Service	Baxley GA	1-800-423-3885
Recycling Specialist	Northfield MN	1-800-742-3558
Refrigerant Reclaim		1-800-238-5902
Refrigerator Recycling	Brunswick OH	1-800-653-6500
Refuse Industry Productions Inc.	Grass Valley CA	1-800-576-3092
Rehrig Pacific	Raymond NH	1-800-882-7440
Rehrig Pacific Co	Erie PA	1-800-458-0403
Resource Technology Group Inc	Northford CT	1-800-242-2801
Reynolds Recycling Hotline		1-800-228-2525
Reynolds Urethane Recycling, Inc	Middleton WI	1-800-858-4244
RGF Environmental Systems	W Palm Beach FL	1-800-842-7771
Road Machinery Sales Inc.	Avon Lake OH	1-800-835-4701
RSI	Glassboro NJ	1-800-533-9977
Rumpke Recycling	Circleville OH	1-800-437-0329
Rural Collection and Tire Recycling	Many LA	1-800-221-1622
Safe Tire Disposal Corp	Cleveland TX	1-800-354-6276
Safe Tire Disposal Corp of Texas	Ardmore OK	1-800-621-2613
Safety Clean	Elgin IL	1-800-323-5740
Sajaw's Tires Recycle	Samson AL	1-800-637-2529
Schaffer Industries	Mahonoy City PA	1-800-544-2006
Scrap-It	Canby OR	1-800-321-7434
Selco Products	Baxley GA	1-800-447-3526
Soil Recycling Technologies Inc.	Hunt Valley MD	1-800-522-7645
Solvent Recovery System	Houston TX	1-800-367-5773
Sonoco Fibre Drum	Marietta GA	1-800-362-3786
Spar Co	Fairfield OH	1-800-624-2069

Steel Recycling Institute		1-800-YESICAN
Swift & McCormack	Redmond OR	1-800-992-8864
The Worm Concern	Thousand Oaks CA	1-800-854-1244
Tin Man Inc.	Chattaroy WA	1-800-447-1076
Tire Management	Miamisburg OH	1-800-848-4990
Tire Shredders	Hudson FL	1-800-992-9359
Toter Inc	Statesville NC	1-800-424-0422
Toter Inc	Statesville NC	1-800-772-0071
Toter Inc.	Statesville NC	1-800-424-0422
Travilah	Rockville MD	1-800-772-1551
Tri State Recycling Services	Dubuque IA	1-800-524-2747
Ultra Image Technologies	Buffalo NY	1-800-851-1276
United Auto Recyclers	Hopkinsville KY	1-800-242-5330
United Metal Receptacle	Pottsville PA	1-800-233-0314
United Recyclers	Suffern NY	1-800-232-7005
US Tire Recycling LP	Concord NC	1-800-328-8473
USF Insulation	Delphos OH	1-800-232-6865
Usher Oil Co	Detroit MI	1-800-874-3730
Wabash Alloys	Cleveland OH	1-800-321-0469
Waste Management of Tri City	Kingsport TN	1-800-554-3098
Wayne Enterprises	Berrien Springs MI	1-800-543-7699
Wheelabrator Environmental Systems	Hampton NH	1-800-682-0026
Wildcat Composting		1-800-627-3954
Willett Oil Co	Des Moines IA	1-800-382-2744
Windsor Barrel Works	Kempton PA	1-800-527-7848
Witte Sanitation	Glencoe MN	1-800-809-4883
Worm Concern The	Simi Valley CA	1-800-854-1244

RISK MANAGEMENT

Accredited Risk Management Specialists	Dallas TX	1-800-348-4866

SAFETY PRODUCTS & SERVICES

Abanda	Decatur AL	1-800-736-2628
Action Ergonomics	Raleigh NC	1-800-482-3587
Alpha-Med technologies Inc.	Cumberland RI	1-800-832-4087
American Safety & Abatement	St. Louis MO	1-800-365-8535
Amotek/Lusa Inc.	Bridgeport CT	1-800-242-4777
H. L. Bouton Co. Inc.	Buzzards Bay MA	1-800-426-1881
Cabot Safety Products	Southbridge MA	1-800-225-9038
Caretek		1-800-997-1994
Champion Ergonomics	Wake Forest NC	1-800-243-1803
Chattanooga Group Inc.	Hixson TN	1-800-592-7329
CMI Inc.	Owensboro KY	1-800-835-0690
Columbus McKinnon Corp	Amherst NY	1-800-888-0985
Comasec Safety Inc.	Enfield CT	1-800-333-0219
Comco Environmental and Safety Services	Bellflower CA	1-800-327-5570
Control Resource Systems Inc.		1-800-272-3786
Direct Safety	Phoenix AZ	1-800-528-7405
DNV Loss Control Management	Loganville GA	1-800-486-4524
Fisher Scientific	Pittsburgh PA	1-800-766-7000
Global Occupational Safety	Port Washington NY	1-800-336-3818
Heppner Risk Management		1-800-662-8469
K-10 Inc.	Mission TX	1-800-531-7496
Kelley Technical Coatings Inc.	Louisville KY	1-800-458-2842

KEM Medical Products Corp.	Farmingdale NY	1-800-553-0330
LIFE Corp.	Milwaukee WI	1-800-700-0202
Manus Products-Florida Inc.	Naples FL	1-800-326-2687
Modern Safety Techniques	Hicksville OH	1-800-542-6646
Neotronics N.A.	Gainesville GA	1-800-535-0606
Norcross Footwear Inc. Safety Products	Louisville KY	1-800-777-9021
NUS Training Corp.	Gaithersburg MD	1-800-338-1505
Ortho-mold	Brookline MA	1-800-344-1012
PGI Inc.	Green Lake WI	1-800-558-8290
Rehab Plus Safety Products Inc.	Lubbock TX	1-800-288-8059
Safeguard Technologies	Leesport PA	1-800-AIRBELT
Safety Corp.		1-800-359-9462
The Saunders Group	Minneapolis MN	1-800-654-8357
Slip Safe International Inc.	Santa Cruz CA	1-800-927-5662
Steelcase		1-800-333-9939
TCB Inc.	Greensboro NC	1-800-216-1953
Titmus Optical Inc.	Petersburg VA	1-800-446-1802
Trusty-Step International	Lynn MA	1-800-323-0047
Unified Safety Corporation	Beverly Hills CA	1-800-394-5776
Wellsource		1-800-533-9355
Wilson-Greene Associates	Portland ME	1-800-634-4327
ZEKS Air Drier Corp.	Malvern PA	1-800-888-2323

SAFETY TRAINING AND EDUCATION

BNA Communications Inc.	Washington D.C.	1-800-233-6067
Comprehensive Loss Management Inc.		1-800-533-2767
Don Brown Productions	Orange CA	1-800-359-7910
Du Pont Safety & Environmental Resources	Philadelphia PA	1-800-532-SAFE
ELB & Associates	Chapel Hill NC	1-800-334-5478
Executive Enterprises	Washington DC	1-800-831-8333
New Environment Inc.		1-800-732-3073
OSHA Training Institute	San Diego CA	1-800-358-9206
Quality Safety Services Inc.	San Antonio TX	1-800-259-0016
Rust Environmental & Infrastructure		1-800-868-0373
Safety Training Systems	Portland OR	1-800-537-8352
The Safety Center	Sacramento CA	1-800-825-7262
Scaggold Training Institute	Houston TX	1-800-428-0162
Visucom Systems	Redwood City CA	1-800-222-4002
Woodward Clyde		1-800-552-9953
R.V. Yale Associates	Seabeck WA	1-800-845-8065

SCHOOLS & UNIVERSITIES - ENVIRONMENTAL, HAZARDOUS MATERIALS & WASTE MANAGEMENT

Berry College	Rome GA	1-800-682-3779
Berry College	Rome GA	1-800-237-7942
Boise State University	Boise ID	1-800-824-7017
Clemson Univerisity		1-800-443-6392
Cornell University	Ithaca NY	1-800-457-3382
Depaul University	Chicago IL	1-800-433-7285
Eagle Mountain Outpost	Sandpoint ID	1-800-654-0307
University of Findlay	Findlay OH	1-800-521-1292

George Washington University	Washington DC	1-800-424-9773
Harvard University	Allston MA	1-800-248-1878
Kansas State University	Manhattan KS	1-800-232-0133
Loma Linda University	Riverside CA	1-800-874-5587
McNeese State University	Lake Charles LA	1-800-622-3352
Montana College of Mineral Science	Butte MT	1-800-445-8324
Outward Bound USA	Greenwich CT	1-800-243-8520
University of Pennsylvania	Edinboro PA	1-800-626-2203
Purdue University	Westville IN	1-800-872-1231
Rutgers University	New Brunswick NJ	1-800-447-8843
University of Tennessee	Knoxville TN	1-800-343-6397
Texas A & M	College Station TX	1-800-826-8331
Texas Tech University	Amarillo TX	1-800-835-5147
University of Houston	Houston TX	1-800-247-7184
Washington State University	Pullman WA	1-800-468-8387
Western Michigan University	Kalamazoo MI	1-800-882-9818

SOFTWARE - ENVIRONMENTAL, HEALTH, SAFETY & GIS

ACS Software	Houston TX	1-800-227-5558
Aldritch Chemical Co.	Milwaukee WI	1-800-558-9160
Anspec Co. Inc.	Ann Arbor MI	1-800-521-1720
BJ Software Systems	Friendswood TX	1-800-771-3007
J. T Baker Inc.	Phillipsburg NJ	1-800-582-2537
Behavior Science Technology Inc.		1-800-548-5781
Bioscience Inc.	Bethlehem PA	1-800-627-3069
C Alexander & Associates Inc.		1-800-433-3761
Camile Products	Midland MI	1-800-252-1977
Chemtox System		1-800-338-2815
Chesapeake Software	Chadds Ford PA	1-800-229-6813
Consolve	Lexington MA	1-800-241-2431
Ctek		1-800-344-LIFT
Digicolor Inc.	Columbus OH	1-800-848-6448
Donlee Technologies	York PA	1-800-366-5334
Dranetz Technologies Inc.	Edison NJ	1-800DRANTEC
EIS International	Rockville MD	1-800-999-5009
EM Science	Gibbstown NJ	1-800-222-0342
ES Data Inc.	Missouri City TX	1-800-372-3282
Enercon Services Inc.	Tulsa OK	1-800-735-7693
Enviro Audit Phase I Software	Indiatlantic FL	1-800-365-3962
EnviroBase Systems Inc.	New Castle DE	1-800-355-9136
Enviromatix Chesapeake Software Div	Chadds Ford PA	1-800-229-5813
Envirometrics	New Castle DE	1-800-355-9136
Environmental Data Network	Williamsville NY	1-800-209-9990
Environmental Data Resources Inc.	Southport Ct	1-800-352-0050
Environmental Risk Information and Imaging	Alexandria VA	1-800-989-0402
Environmental Systems Corp		1-800-688-7900
ERM Enflex Data	Philadelphia PA	1-800-544-3117
Fisons Instruments	Beverly MA	1-800-999-5011
Genium Publishing Corp.	Schenectady NY	1-800-243-6486
Geocomp Corp.	Concord MA	1-800-822-2669
Geosoft Inc.	Toronto ON CN	1-800-363-6277
Golden Software Inc.	Golden CO	1-800-972-1021

HRD Software	Amherst MA	1-800-822-2801
Hazco Services	Dayton OH	1-800-332-0435
Heart Information Systems Inc.	Schenectady NY	1-800-572-6501
Hewlett Packard Co.	Palo Alto CA	1-800-227-9770
I-Chem	Hayward CA	1-800-443-1689
I-Chem		1-800-262-5006
Intergraph Corp.	Huntsville AL	1-800-345-4856
IPT Corp.	Palo Alto CA	1-800-944-5468
Isco Inc.	Lincoln NE	1-800-228-4373
IT Corp.	Wilmington CA	1-800-421-5574
IT Corp.		1-800-824-7932
IT Corp.		1-800-262-1900
JB Systems Inc.	Woodland Hills CA	1-800-275-5277
Jordan Systems Inc.	Cedar Rapids IA	1-800-859-3023
Leupold & Stevens	Beaverton OR	1-800-452-5272
Logical Technology Inc.	Peoria IL	1-800-3-PEORIA
Meteorological Evaluation Services	Amityville NY	1-800-952-2052
Micro Dataware	Hatward CA	1-800-582-6624
MicroCal Software	Northampton MA	1-800-969-7720
Micromedex	Denver CO	1-800-525-9083
Mitchell Scientific	Westfield NJ	1-800-537-9779
National Instruments	Austin TX	1-800-433-3488
North American Software	Tustin CA	1-800-966-5678
OSHA-Soft Inc.	Amherst NH	1-800-446-3427
Pavillion Technologies		1-800-880-5432
Permit Tracker	Indialantic FL	1-800-365-3962
Pico Info Systems	Springfield OR	1-800-488-7823
Pro-Am Software	Warrendale PA	1-800-852-7316
Regscan Technology Corp.	Williamsport PA	1-800-326-9303
Safety Software Inc.	Charlottesville VA	1-800-932-9457
Safeware Inc.	San Mateo CA	1-800-229-9400
SampleTrak		1-800-499-9505
Savant Audiovisuals Inc.	Fullerton CA	1-800-472-8268
Sierra Geophysics Inc.	Seattle WA	1-800-826-7644
Sigma Chemical Co.	St Louis MO	1-800-325-3010
Simon Hydro-Search	Golden CO	1-800-544-5528
Simon Hydro-Search	Houston TX	1-800-548-7667
Simon Hydro-Search	Irvine CA	1-800-882-8123
Systems & Software Group	Ann Arbor MI	1-800-860-SSG1
Utilicom Inc.	Pittsford NY	1-800-926-6759
Versar Inc.	Springfield VA	1-800-283-7727
Virtual Media Corp	Scottsdale AZ	1-800-645-4130
Vista Environmental Information	San Diego CA	1-800-733-7606
YSI Inc.	Yellow Springs OH	1-800-765-4974

SOFTWARE - SAFETY

Aldrich Chemical Co. Inc.		1-800-231-8327
Behavior Science Technology Inc.		1-800-548-5781
C Alexander & Associates Inc.		1-800-433-3761
Chemgrate		1-800-345-5636
The Chemtox System		1-800-338-2815
CHR Systems Group Inc.		1-800-966DATA
Corbus Inc.		1-800-524-7096
Ctek		1-800-344-LIFT
Genium Publishing Corp.		1-800-243-6486
IHS	Englewood CO	1-800-241-7824

Infection Control & Prevention Analysts Inc.		1-800-426-8015
Logical Technology	Peoria IL	1-800-3-PEORIA
Occupational Health Services Inc.		1-800-445-6737
OSHA-Soft Inc.		1-800-446-3427
Parsec Inc.		1-800-527-3454
Pico Info Systems	Springfield OR	1-800-488-7823
PRO-AM Software		1-800-852-7316
Regulation Scanning		1-800-326-9303
Safety Software Inc.	Charlottesville VA	1-800-932-9457
Systems & Software Group	Ann Arbor MI	1-800-860-SSG1

TIMBER & TIMBERLAND MANAGEMENT

Atlantic Timber	Madison Heights VA	1-800-344-0947
Baker Timber	Fuquay Varina NC	1-800-552-4965
Brady Timber Inc.	Mt Enterprise TX	1-800-874-6177
Burney FH	White Oak NC	1-800-452-1209
Canal Wood of Lumberton	Lumberton NC	1-800-782-0450
Canal Wood of Lumberton	Lumberton NC	1-800-872-2647
Carnet Potter	Plainview AR	1-800-451-0422
Columbia Timber	Gainesville FL	1-800-344-1473
Columbia Timber Co. Inc.	Magnolia AR	1-800-356-3979
Delaware Pulpwood	Lincoln DE	1-800-441-8111
Eastex Wood Products	Jasper TX	1-800-222-3451
Florida Forestry Services		1-800-338-0567
Forestry Exports USA	Kingwood TX	1-800-638-9119
Greenville Timber Corp	Madison FL	1-800-533-4902
Idaho Timber	Halstead KS	1-800-572-9663
Idaho Timber Corp	Boise ID	1-800-654-8110
Illinois Forestry Service		1-800-524-6063
Irv Morse Sales Inc.	W Babylon NY	1-800-223-0523
Lowe Timber	Huntsville TX	1-800-952-8328
Mark King Forestry Consultant Service	Winona MS	1-800-858-5464
McCormick Gerald Sawmill	Fountain MI	1-800-858-6455
MGM Logging	Pineland TX	1-800-232-9640
Pine State Chipping	Stoneville NC	1-800-227-2488
Remington Southeastern Inc.	Forsyth GA	1-800-732-3858
Rhodes Forestry Services Inc.	Aiken SC	1-800-348-5263
Stallworth Timber	Beatrice AL	1-800-633-6862
Stone Container Corp	Columbia SC	1-800-438-0412
Texas Land & Timber Co. Inc.	Buna TX	1-800-231-8124
Timber Harvester	New Bern NC	1-800-527-0735
Trowbridge Forest Products	Hampton CT	1-800-367-3785
Vanderwal Jan	Longview TX	1-800-328-8804
WI Forest Products	Bonners Ferry ID	1-800-548-8221
Weatherford Alex	Diboll TX	1-800-528-1679

TRAINING AND EDUCATION - HAZARDOUS MATERIALS, REGULATORY & SAFETY

BNA Communications	Rockville MD	1-800-233-6067
CAE Seminars	Palatine IL	1-800-223-7364
CAS Registry Services	Columbus OH	1-800-848-6538
Cambridge Information Group	Bethesda MD	1-800-843-7751
Coastal Video Communications	Virginia Beach VA	1-800-767-7703

Dupont Co. Safety & Environmental Res's	Wilmington DE	1-800-532-7233
Emergency Film Group	Plymouth MA	1-800-842-0999
Environmental Resource Center	Cary NC	1-800-537-2372
Genum Publishing Corp	Schenectady NY	1-800-243-6486
Hazmat Training Inc.	Murfreesboro NJ	1-800-868-4871
Hazmat Training, Information and Services	Columbia MD	1-800-777-8474
Haztech International	Bellevue WA	1-800-468-7644
Heath Consultants Inc.	Houston TX	1-800-432-8487
Industrial Training Inc.	Grand Rapids MN	1-800-253-4623
JJ Keller & Associates	Neenah WI	1-800-558-5011
Reactives Management Corp.	Chesapeake VA	1-800-372-6742
Rust Environmental & Infrastructure	Greensville SC	1-800-868-0373

WASTE CONTAINERS & STORAGE

A S Steel Drum Co	Brooklyn NY	1-800-257-2945
Allied Technology Group	Fremont CA	1-800-227-2840
C & S Container	Ft Wayne IN	1-800-419-0404
Consolidated Waste Service	Toms River NJ	1-800-388-2972
International Quality Products	Cypress TX	1-800-556-1456

WASTE MANAGEMENT SERVICES - DISPOSAL, TREATMENT, REDUCTION & RECYCLING

A E T C	Rocky Hill CT	1-800 362-2382
AAD Disposal	Gary IN	1-800-526-8386
ABB Sanitec	Wayne NJ	1-800-551-9897
ABC Containers	Walterboro SC	1-800-562-5079
Able Disposal	Chesterton IN	1-800-828-2253
Acme Waste Systems	Ossian IN	1-800-541-7904
Advanced Environmental Technology Corp	Williamburg VA	1-800-972-7622
Advanced Recycling (Wood)	Newark NJ	1-800-932-4467
Advanced Waste Services Inc	Rockford IL	1-800 842-9792
Advanced Waste Services Inc.	Brown Deer WI	1-800-842-9792
AETC	Houston TX	1-800 887-2382
AETC	Rocky Hill CT	1-800-362-2382
AETC	Houston TX	1-800-262-2382
AFO	Luling TX	1-800-533-5256
AJM Disposal	Philadelphia PA	1-800-453-9314
Alabama Waste Serevices	Anniston AL	1-800-732-7891
All State Power Vac	Linden NJ	1-800-437-3484
All Town & Country Septic Tank & Drain	Barberton OH	1-800-992-8531
Allegro Carting	Hoboken NJ	1-800-542-1748
Alleycat Inc.	Racine WI	1-800-362-1783
Amco Resource Recovery Inc	Hyattsville MD	1-800-544-8619
American Environmental Management	Rancho Cordova CA	1-800 345-2362
American Environmental Management	Sacramento CA	1-800-345-2362
American Hospital Consultant Co.	Macon GA	1-800-634-9106
American MFG	Manassas VA	1-800-345-3132
American Refuse Control Inc.	Jackson GA	1-800-344-2021
American Refuse Systems ARS	Harrisonburg VA	1-800-447-2771
American Rerefining Co-ARCOM	Tacoma WA	1-800 831-5243
American Sanitation	Miramar FL	1-800 238-1426
American Waste & Pollution Control	New Orleans LA	1-800-521-4618

Anaheim Mfg	Anaheim CA	1-800 854-3229
Anco Leasing Corp	Hudson NH	1-800-292-4285
Applied Wste Water Services Inc.	Belle Mead NJ	1-800-334-1230
Aptus	Coffeyville KS	1-800 248-0442
Aptus	Coffeyville KS	1-800 292-2558
Arena Trucking Co Inc	Rice VA	1-800 908-7274
Arkansas Waste Disposal	Little Rock AR	1-800-732-0021
Armor Environmental Services	Columbia TN	1-800 462-7712
Arrow Disposal of Smyrna Inc.	Wilmington DE	1-800-533-9050
ASG Disposal Services	Nashua NH	1-800-258-3783
Atlantic Waste Systems North	Lynn MA	1-800 445-1318
Atlas Refuse Disposal	Crestwood IL	1-800-356-9982
Aurora Cellulose	St. Petersburg FL	1-800-282-1153
B & E Cartage	Lenore WV	1-800 262-1017
Baker Sanitation Inc.	Cortez CO	1-800-537-6731
Bale Press	Douglasville GA	1-800-241-2363
Beachside Services	Panama City FL	1-800-628-7560
Bealine Service Co Inc.	Pasadena TX	1-800-545-4210
Benson Enterprises	Easton MA	1-800-323-3339
Berkshire Pittsfield Septic Cleaning Service	Berkshire MA	1-800-364-8265
Bermans Diversified Industries	San Jose CA	1-800 942-5222
BES Environmental Specialists Inc.	Kingston PA	1-800-451-6527
BFI	Johnston RI	1-800-622-3727
BFI Medical Waste Systems Of Utah	Salt Lake City UT	1-800 321-9993
BFI Niagara Frontier	Sheridan NY	1-800-882-4126
BFI of Northeast Tennessee	Rogersville TN	1-800-551-3747
BFI Waste Systems	Huntington WV	1-800 331-0988
BFI Waste Systems	Geneva NY	1-800-262-2652
BFI Waste Systems	Huntington WV	1-800-331-0988
BFI Waste Systems Inc.	Muskego WI	1-800-232-0860
BFI Waste Systems Inc.	Amarillo TX	1-800-692-4436
Big Dipper Enterprises Inc.	Wahpeton ND	1-800-654-4306
Bio Environmental Services Inc.	Bluefield WV	1-800-525-7602
Bio Hazard Management	Tyler TX	1-800 722-0117
Bio Nomic Service	Charlotte NC	1-800-782-6798
Bio-Hazard Management	Tyler TX	1-800-722-0117
Bio-Safe America	Bradenton FL	1-800-842-3912
Biomedical Service Corp	Lake City GA	1-800-526-4864
Bowers Sanitation	Vickery OH	1-800 223-8727
Brian's Waste Handling	Lumberton NC	1-800-238-0063
Browning Ferris Industries	S Yarmouth MA	1-800-352-7808
Browning Ferris Industries	Brockton MA	1-800-843-1666
Browning Ferris Industries	Marshfield MA	1-800-843-2234
Browning Ferris Industries	Woburn MA	1-800-334-3126
Browning Ferris Industries	Monroe LA	1-800-222-2925
Browning Ferris Industries		1-800-551-5584
Browning Ferris Industries	Jackson MS	1-800-635-1258
Browning Ferris Industries	Marion IL	1-800-634-0215
Browning Ferris Industries	Kersey PA	1-800-257-5705
Browning Ferris Industries	Hagerstown MD	1-800-952-9888
Browning Ferris Industries	Fall River MA	1-800 843-1666
Browning Ferris Industries	Houston TX	1-800 338-7814
Browning Ferris Industries	South Yarmouth MA	1-800 352-7808
Browning Ferris Industries Inc.	Chicopee MA	1-800-367-7778
Browning Ferris Industries Medical Waste	Ft Lauderdale FL	1-800 437-4447
Browning Ferris Industries of Mississippi	Biloxi MS	1-800-443-6562

Browning Ferris Industries of Mississippi	Greenville MS	1-800-523-6437
Browning Ferris Industries Waste Systems	Tyngsboro MA	1-800 238-9020
Browning-Ferris Industries of Mississippi	Greenville MS	1-800 523-6437
Bryan Edward Sons Ltd	E Dedham MA	1-800 338-3867
BSE Recycling Works Corp	Windham NH	1-800-639-6306
Budoff Iron & Metal Inc.	Akron OH	1-800-523-7193
Burney the Burner	San Bernadino CA	1-800-272-1243
Byers Recycling Disposal Facility	Logansport IN	1-800-342-9377
C & H Sewerage Service	Elmer NJ	1-800 367-0147
C & M Co.	Winston-Salem NC	1-800-262-9762
Cadence Chemical Resources Inc.	Michigan City IN	1-800-866-0688
Caldeira Brothers	Waretown NJ	1-800-732-3259
Cardella Trucking	N Bergen NJ	1-800-548-7001
CDS of Charleston	Goose Creek SC	1-800-521-1797
Chem Met Services Inc.	Wyandotte MI	1-800-282-9251
Chemfix Technologies	Metaire LA	1-800-992-3443
Chemical Waste Management	Conley GA	1-800 843-3604
Chemical Waste Management	Azusa CA	1-800-541-8516
Chemical Waste Management	Conley GA	1-800-232-8947
Chemical Waste Management	Houston TX	1-800-445-3589
Chemical Waste Management	Oak Brook IL	1-800-843-3604
Chemical Waste Management	Princeton NJ	1-800-533-4133
City Industrial Cyclone	Harbor City CA	1-800-642-5228
Clatskanie Sanitary Service	Clatskanie OR	1-800-422-9998
Clean City Squares	St. Louis MO	1-800-231-1327
Clean Harbors of Maine	S Portland ME	1-800-526-9191
Clearwater Sales Inc.	Marshall TX	1-800-635-5864
Coate Septic Tank	W Milton OH	1-800-762-4066
Colorado Incineration Services Inc.	Denver CO	1-800-221-6456
Commercial Burning Systems Inc.	Wheeling IL	1-800-247-1016
Computerized Waste Systems	Louisville KY	1-800-233-9923
Connecticut Carting	Franklin CT	1-800-654-5799
Conservation Services	Bennett CO	1-800-662-4089
Consolidated Bailing Machine Co. Inc.	Jacksonville FL	1-800-874-8328
Consolidated Waste Service	Hamilton NJ	1-800 514-0061
Continental Vanguard	Bellmawr NJ	1-800-223-4525
Cousins Waste Control	Toledo OH	1-800-433-6754
Crestwood Waste Recovery Systems Inc	Newark NJ	1-800 622-4446
Crestwood Waste Recovery Systems Inc.	Saddlebrook NJ	1-800-622-4446
CSI Of Northern Kentucky	Boone County KY	1-800 847-8520
Cuesink	Orlando FL	1-800-327-7791
Curry Transfer and Recycling	Gold Beach OR	1-800-826-9801
D'Ambra Construction Co.	Warwick RI	1-800-966-SOIL
Daubs Disposal	Browns IL	1-800-892-7838
Davis and Davis Waste Haulers	Florence SC	1-800-327-4562
Davis Landfill and Sanitation	Tonkawa OK	1-800-522-6437
Deffenbaugh Disposal Service	Shawnee KS	1-800 631-3301
Delvecchio Transport Material Handling	Dunmore PA	1-800-272-4020
Dependable Disposal Hauling Co	Lancaster CA	1-800 514-4095
Dependable Environmental Services Inc.	Windham NH	1-800-426-3139
Dickson Equip Co. Inc.	Dallas TX	1-800-634-7531
Disposal Control Services	Albuquerque NM	1-800-433-4981
Disposal Systems	Tom Bean TX	1-800 328-9455
Disposal Systems	Bells TX	1-800-328-8455
DMP Corp	Rock Hill SC	1-800 548-5614
Douglas Disposal	S Lake Tahoe CA	1-800-356-1242

Dynamerican	Copley OH	1-800 362-8490
E & T Shredding Protection Inc.	Jacksonville FL	1-800-343-5755
E W R Inc	Coal City IL	1-800 441-7358
Eastern Chemical Waste Systems	Silver Springs MD	1-800 826-0518
Eastern Chemical Waste Systems Inc.	Washington DC	1-800-255-3954
Eastern Chemical Waste Systems Inc.	Silver Springs MD	1-800-826-0518
Eastern Ecology Services	Bronx NY	1-800-232-4568
Eastern Pump Supply Inc.	Danbury CT	1-800-486-1107
Eastern Waste Ind.	Greensboro MD	1-800-492-5999
Eco Safe Recycling	Orlando FL	1-800-522-4326
Eco Systems	St Louis MO	1-800-325-3007
Ecological Labs	Freeport NY	1-800 645-2976
Ecological Labs	Freeport NY	1-800-645-2976
Ecological Systems	Indianapolis IN	1-800 235-1661
Ecologix	Wellman IA	1-800-433-2999
Ecology Equip	Pittsburgh PA	1-800-852-0094
Ecotec	Litchfield ME	1-800-338-2164
Ecova	Redmond WA	1-800-548-3668
Ecova Corp	Kimball NE	1-800 462-4678
Ecoworld International	Rancho Mirage CA	1-800-538-2117
Ed's All Clean Disposal Inc.	Charleston WV	1-800-345-4660
Edward Byan Sons Ltd.	E Dedham MA	1-800-338-3867
EGH Disposal	Elma WA	1-800-345-1529
EGI Inc.	Fair Oaks CA	1-800-633-9783
Elda Recycling & Disposal Facility	Cincinnati OH	1-800-238-3532
Emtec	Irvine CA	1-800-438-3683
Envirite Corp.	Pontiac IL	1-800-348-3425
Envirite Corporation	Plymouth Meeting PA	1-800 858-9423
Enviro Med Associates Inc	Wilkes-Barre PA	1-800 673-2511
Enviromed Inc.	Meridian MS	1-800-832-6999
Environment Technology Inc	Crossville TN	1-800 634-8649
Environmental Contracting and Supply	La Grange IL	1-800-331-1945
Environmental Logistics Inc.	Houston TX	1-800-227-2850
Environmental Options	Houston TX	1-800-346-4606
Environmental Systems	Albuquerque NM	1-800-732-2002
Environmental Technology Inc.	Crossville TN	1-800-634-8649
Environmental Waste Disposal Inc.	Huntington WV	1-800-522-1578
Environmental Waste Management	Mohnton PA	1-800 638-6239
Environmental Waste Reduction	Chamblee GA	1-800 754-8441
Environmental Waste Resources Inc.	Waterbury CT	1-800-225-5397
Envirosafe Services Of Ohio	Toledo OH	1-800 537-0426
Envirosave USA	Carolina Beach NC	1-800 562-8869
Envirotec Enterprises Inc	Tucson AZ	1-800 647-6832
Envirotech of America	E Syracuse NY	1-800-448-3851
Envirovac Inc.	Rockford IL	1-800-435-6951
Envotech Management Services Inc.	Belleville MI	1-800-368-6835
ESI Rogers Oil	Waterloo WI	1-800-822-9608
Essex Waste Management Service Inc.		1-800 966-6465
EWR Inc.	Coal City IL	1-800-441-7358
Falcon Disposal Service	Wilmington CA	1-800-252-1211
Federal Environmental Services Inc.	Roswell GA	1-800-635-3009
Federal Environmental Services Inc.	Walterboro SC	1-800-637-6693
Fergus Power Pump Inc.	Fergus Falls MN	1-800-243-7584
Filter Flat Of Ohio	Milford OH	1-800 888-3528
Fine Surgical Inc.	Lynbrook NY	1-800-851-5155
Finish Engineering Co.	Kensmore NY	1-800-336-2550

Floyd's of South Carolina Inc.	Florence SC	1-800-992-0274
FM Systems Inc	Golden CO	1-800 367-9786
Fox Equip	Shirley NY	1-800 635-6801
Fuel Processors Inc.	Portland OR	1-800-367-8894
Georgia Disposal	Thomson GA	1-800-841-4388
Geowaste Of Ga	Valdosta GA	1-800 238-1019
GFS Technologies	Cleveland TN	1-800-882-4378
Good Sense Company	Telluride CO	1-800-462-4042
Government Innovators Inc.	Phoenix AZ	1-800-528-5308
Grand Central Sanitation Inc.	Pen Argyl PA	1-800-621-2100
Gravity Flow Systems	Carbondale PA	1-800-237-7500
Great Lakes Environmental Services	Warren MI	1-800-451-9782
Green Alternatives	San Jose CA	1-800-345-3363
Greens Disposal Service	Great Falls MT	1-800-257-6947
GSX Chemical Services Inc.	Laurel MD	1-800-638-4440
GSX of Greater Boston	Charlestown MA	1-800-654-4496
GSX Services (Laidlaw)	Greenbrier TN	1-800-251-1227
GSX Services Inc	Clearwater FL	1-800 421-0553
GSX Services Of California	Martinez CA	1-800 445-0203
H. B. Antifreeze Recycling Service	Columbia PA	1-800-437-0262
Hamaking Corp.	Acton MA	1-800-597-3342
Hamilton County Landfill	Webster City IA	1-800 535-1145
Hampden Color & Chemical	Springfield MA	1-800-225-3546
Hatchers Disposal Inc.	Hudson FL	1-800-282-9820
Herrick Valley Recycling & Disposal Facility	Adena OH	1-800-367-1294
Highway 36 LDC	Aurora CO	1-800-392-3636
Hitemp Inc.	Cleveland OH	1-800-242-4422
Huco Environmental Service	Jackson MI	1-800 475-8343
Hukill Chemical Corp	Cleveland OH	1-800-962-1143
Hupp Concrete Products	Zanesville OH	1-800-282-9519
Hydro Chem Service	San Francisco CA	1-800-822-0019
Incendere	Norfolk VA	1-800-872-2876
Independent Rubbish Service Inc.		1-800-841-0248
Industrial Lubricants Hazardous Waste Mgt	Tacoma WA	1-800 582-2343
Industrial Waste Disposal Co Inc	Dayton OH	1-800 228-1336
Inskip Glass Recycling Inc.	Knoxville TN	1-800-922-4527
J & F Disposal	Carrollton MO	1-800-626-5136
J & J Baker Enterprises Inc	Punta Gorda FL	1-800 562-9599
J P Mascaro Sons Inc.	Harleysville PA	1-800-368-6895
J W C Environmental	Norcross GA	1-800 762-9593
Jackson Septic Tank & Construction Co Inc		1-800-922-7430
JCI Environmental Services	Los Angeles CA	1-800 336-8055
Jewell Disposal Service Inc.	Lebanon TN	1-800-842-3842
Jim Wilson Distributing	Salem OH	1-800-422-9276
Joe's Rescue White Pass City Sanitary	Centralia WA	1-800-525-4167
JWC Environmental	Lancaster PA	1-800-762-9593
Kabco	Raleigh NC	1-800-532-1370
Kahle Hauling	Owensville MO	1-800-435-2379
Karney and Associates	Newrak Valley NY	1-800-227-0733
Ken's Pick-Up Service Inc.	Traverse City MI	1-800-238-8577
Kenetech Resource Recovery	Sarasota FL	1-800 741-4439
Kidd Septic Tank	Myrtle Beach SC	1-800 634-5775
Kinsbursky Brothers Inc	Anaheim CA	1-800 548-8797
Laidlaw Environmental Service	Buttonwillow CA	1-800 544-7199
Laidlaw Environmental Services	McKittrick CA	1-800-544-7199
Laidlaw Environmental Services	Reidsville NC	1-800-334-5953

Laidlaw Environmental Services	Columbia SC	1-800-356-8570
Laidlaw Environmental Services	Columbia SC	1-800 465-2435
Laidlaw Waste Systems	Liverpool NY	1-800-428-8492
Laidlaw Waste Systems	Edwardsville IL	1-800-222-5158
Laidlaw Waste Systems	Rowley MA	1-800-233-0309
Laidlaw Waste Systems		1-800-892-0292
Laidlaw Waste Systems		1-800-872-0208
Lake Area Disposal	Rice Lake	1-800-542-2124
Lake Area Disposal Landfill	Sarona WI	1-800-847-3396
Lefco Environmental Technology	Montgomery TX	1-800-533-2688
Lewis Farms & Liquid Waste Removal Inc.	Burgaw NC	1-800-624-2979
Lodal Inc.	Kingsford MI	1-800-435-3500
Longview Waste Systems	Canton MO	1-800-874-4612
Lundell Mfg	Cherokee IA	1-800-352-4639
M & P Enterprises	Huntington UT	1-800-822-3400
M & T Services Inc.	Gaffney SC	1-800-255-6176
Madison Disposal	Waco KY	1-800-331-9241
Magnum Tank Service	Pompano Beach FL	1-800-235-0189
Maine Biomedical Service	Springvale ME	1-800-438-0801
Maintenance Management	Hazlet NJ	1-800-242-2547
Marathon Equipment Co.	Yerington NV	1-800-624-5724
MBI Industries	Monroe WA	1-800-845-7404
Medical Disposal Service Inc	St Petersburg FL	1-800 521-4541
Medical Disposal Systems Inc	Burnsville MN	1-800 235-0137
Medisin	Prestonsburg KY	1-800 822-8012
Michelin Tire	Sherbourne MA	1-800-422-1682
Michigan Disposal Service	Kalamazoo MI	1-800-221-7191
Mid Atlantic N-Viro	Hatfield MA	1-800-247-9316
Mid Atlantic Oil Management Service Inc.	Baltimore MD	1-800 331-5408
Midwest Medical Disposal	Owensville MO	1-800 874-2057
Midwest Tire Services	Brooklyn Park MN	1-800-428-8892
Mills Garbage Service	Kenova WV	1-800-272-7430
Mineral Springs	Port Washington WI	1-800-932-6216
Missouri Disposal	Galt MO	1-800-346-6844
MKC Enterprises Inc.	Doraville GA	1-800-457-6521
MM Solid Waste Equipment Inc.	Lee NH	1-800-258-7370
Mobile Process Technology	Memphis TN	1-800 239-3028
Mobile Reclamation System	W Sacramento CA	1-800-824-0065
Modern Waste Treatment Systems	Urbana IL	1-800 432-1265
Mondo Bruce & Septic Service	Monroe CT	1-800-262-4765
Montana Superfund Hotline	Helena MT	1-800-648-8465
MTI Waste Removal System	La Grange GA	1-800-537-4990
Municipal Services Corp	Sawyer ND	1-800-554-3289
National Incinerator Service	Corsicana TX	1-800-544-0661
North Star Refuse	Lexington MA	1-800-525-0971
Northeastern Disposal	Montville OH	1-800-238-3348
Northern A1 Services	Kalkaska MI	1-800-544-2663
Northern Business Group	Portland OR	1-800-448-2121
Northwest Enviroservice Inc.	Seattle WA	1-800-441-1090
NU-SOILS	Parsippany NJ	1-800-225-7645
Oasis Oil Filter Washer Crusher	Salt Lake City UT	1-800-722-5202
Oil and Solvent Process Co	Henderson CO	1-800-525-1840
Okanogan Valley Disposal	Oroville WA	1-800-422-0136
Oregon Sanitary Service Institute	Salem OR	1-800-527-7624
Ortho Medical Waste Management	Daytona Beach FL	1-800 633-0868
P & D Biomedical Waste Disposal	Kingman AZ	1-800 551-6248

P&T Container Service	Haverhill MA	1-800-692-0009
Pack Rat Services	Andrews TX	1-800-654-9521
Pactec	Clinton LA	1-800-272-2832
Palm Beach Roll'Off	Boca Raton FL	1-800-522-0477
Parts Inc.	Piedmont SC	1-800-435-7278
Pasco Lakes Wesley	Chapel FL	1-800-352-6991
Pawnee Products	Goddard FL	1-800-533-2475
PCB Ballast Disposal By Eastern Env	Port Chester NY	1-800 808-7227
Pederson Sanitation Corp	Ft Dodge FL	1-800-428-5546
Pegasus Waste Management	Woodburn OR	1-800-354-9033
Penas Disposal	Orosi CA	1-800-535-4004
Peninsula Dump All	Sturgeon Bay WI	1-800-438-8628
Petroleum Management Services Inc.	Reading MA	1-800-932-3057
PFS Waste Control Service	Oildale CA	1-800-262-0321
PGL Group Port	Clinton PA	1-800-231-1798
PGS Carting Inc.	Copaigue NY	1-800-523-4324
Pierce County Refuse Service	Tacoma WA	1-800-345-3629
PMT Services	Baton Rouge LA	1-800-621-4540
Poly Bac	Bethlehem PA	1-800-523-9385
Poor Richards	St Paul MN	1-800-732-5286
Premier Resources	Industry CA	1-800-445-1079
Pure Solve Inc.	Sulphur LA	1-800-472-6074
Quickclean Services	Montague TX	1-800-638-2183
R G & Associates	Conroe TX	1-800 324-6596
Radium Petroleum	Wichita KS	1-800-572-2641
Raff's	Cape May NJ	1-800-272-3320
Rail Container Services Inc	Westchester PA	1-800 852-2494
Rail Container Services Inc.	W Chester PA	1-800-852-2494
RAM Sales	Voorhees NJ	1-800-334-2275
Recontek	Newman IL	1-800 867-8682
Redline Environmental Corp.	Cleveland OH	1-800-752-2430
RedWing Disposal/Port-O-Let	Homosassa Springs FL	1-800-223-4825
RedWing Environmental Technologies	Framingham MA	1-800-498-9870
Refuse Management Systems	Floral Park NY	1-800-346-5926
Refuse Managment Inc	Springfield MA	1-800 440-2001
Refuse Removal Systems Inc.	Fair Oaks CA	1-800-231-2212
Reliable Disposal Service	Edenton NC	1-800-453-8997
Renewable Oil Services Inc.	Kilgore TX	1-800-252-5768
Resource Control	Barre MA	1-800-441-4198
Resource Protection Inc.	Hobbs NM	1-800-328-7801
Robinson Pipe Cleaning	Newark NJ	1-800-242-7257
Roche & Sons Disposal	Layton UT	1-800 772-0273
Rocky Mountain Welding & Fabricating	Pleasant Grove UT	1-800-626-6949
RS Andrwes Waste Management	Tucker GA	1-800-448-8361
Rubbish Removal of New York	Syracuse NY	1-800-562-9920
Rudco Products Inc.	Vineland NJ	1-800-828-2234
Rumke Container Service Inc.	Dayton OH	1-800-543-0477
Rumpke Of Indiana	Orleans IN	1-800 428-6287
Rumpke of Indiana	Bloomington IN	1-800-722-4090
Rumpke Waste	Greenville OH	1-800-752-1887
Safety First Medical Waste Management Inc.	BRONX NY	1-800 432-3241
Safety Storage Co Inc	Hollister CA	1-800 344-6539
Safety-Kleen Corp.	Elgin IL	1-800 323-5740
Salvajor	Kansas City MO	1-800-821-3136
Sanitary Commercial Services	Jackson OH	1-800-752-6760
Sanitary Liquid Waste	Hickory NC	1-800 541-8265

Sanitary Services	Manchester CT	1-800-458-7274
Schofield Inc.	Wayland MA	1-800-545-2275
Seneca Disposal Inc.	Tifflin OH	1-800-447-4030
Service Inc.	Phoenix AZ	1-800-433-4981
Sidener Supply Co.	Columbia MO	1-800-392-7211
Smith's Oil Service	Romulus MI	1-800-354-8563
Solid Waste Composting Council	Washington DC	1-800-457-4474
Solid Waste Management Systems	Pittsburgh PA	1-800-433-7967
Solid Waste Technologies	Warren NJ	1-800-548-9789
Solvent Systems International	Elgin IL	1-800-835-5774
Son Mar Sanitation	Medford NY	1-800 235-0738
Son Mar South Sanitation	W Palm Beach FL	1-800 345-3826
Soresi Chemical Group Inc	Silver Spring MD	1-800 654-9967
South and South Inc.	Columbia TN	1-800-824-2788
Southern Equipment	Amelia Island Plantation FL	1-800-523-3861
Southern Waste Information Exchange	Tallahassee FL	1-800-441-7949
Southern Waste Management	Dallas TX	1-800-442-3065
Southland Equipment	Bartow FL	1-800-282-1654
Southwest Sunbelt Sanitation	Greenville TX	1-800-441-8434
SP Industries	Hopkins MI	1-800-592-5959
Specialty Metals	Birmingham AL	1-800-435-0363
Spectrum Chemical Solvent and Oil	Norco CA	1-800-843-9238
Stamco Environmental	San Martin CA	1-800-321-1030
Suburbia Systems Corp.	Wilkes Barre PA	1-800-782-3384
Suffolk Services Hazardous Waste	Lowell MA	1-800-342-3515
Super Garbage Service	Dalton GA	1-800-382-4714
Superior Sanitation Service	Copperas Cove TX	1-800-852-9716
Target Compaction Inc.	Canastota NY	1-800-433-0828
Taylor Garbage Service	Vestal NY	1-800-223-7001
TCT	S Salem OH	1-800-356-5235
Technical Environmental Systems	La Porte TX	1-800-446-5777
Technical Transporters Inc.	Deer Park TX	1-800-962-1662
Texas Industrial Disposal	Mt. Pleasant TX	1-800-828-8434
Texas Trash Haulers	Whitney TX	1-800-848-9863
Therm/Tec Destruction Services	Sherwood OR	1-800-626-9951
Tisdales Sanitary Service	AUSTIN TX	1-800 543-6501
TMI Equipment Inc.	Houston TX	1-800-248-0514
Tobacco Valley Sanitation	S Windsor CT	1-800-346-3591
Tolman Environmental Services	Baldwinville MA	1-800-231-4873
Total Waste Management Corp	Newton NH	1-800-345-4525
Trashmasters	Forestville MD	1-800-322-6110
Travers Concrete Products	Sunderland MD	1-800-451-9583
Tri City Disposal	Hughes Springs TX	1-800-435-7296
Tri County Waste	Edelstein IL	1-800-782-6705
Tri State Environmental	Southgate MI	1-800-225-9711
Tri-S Inc.	Ellington CT	1-800-828-7471
Tri-S Inc.	Brattleboro VT	1-800-359-3677
Tri-State Medical	Hudson NH	1-800 338-0476
Tricil Environmental Response	Houston TX	1-800-262-0387
Tricil Recovery Services Inc.	Bartow FL	1-800-852-8774
Trinity Environmental Technologies Inc.	Overland Park KS	1-800-722-8378
Triple T Trash Inc.	Carlyle IL	1-800-232-8568
Truman Horner Inc	Harrisburg PA	1-800 255-8479
TTC	S Salem OH	1-800 356-5235
Tuf Tite Septic Components	Barrington IL	1-800-382-7009
Turleys Sanitation Service	Mt Sterling KY	1-800-248-6739

Two Guys and a Truck	Lake Worth FL	1-800-262-5503
United Disposal Service Inc.	Canaan Valley WV	1-800-543-7897
United Waste Removal Service	New York NY	1-800-235-4851
United Waste Systems	Grand Junction CO	1-800 345-9264
United Waste Systems Inc.	Atlanta GA	1-800-382-1677
Universal Waste	Mayfield KY	1-800-457-4679
US Ecology Inc.	Hoffman Estates IL	1-800-852-2055
USA Dano	Portland OR	1-800-872-3266
V & R Trucking	N Arlington NJ	1-800 244-5860
Van Der Molen Disposal (BFI)	Melrose Park IL	1-800-942-0491
Ventura Waste Management	Ventura CA	1-800-952-2909
Vesta Technology Ltd	Ft Lauderdale FL	1-800-545-0690
Von Roll of Ohio	E Liverpool OH	1-800-545-7665
Wastainer	Mukwonago WI	1-800-322-6790
Waste Control	Franklin IN	1-800-624-1611
Waste Control Industries Inc.	Tampa FL	1-800-637-1678
Waste Controls Corp.	Galva IL	1-800-535-9012
Waste Handling Systems	Forest City NC	1-800-626-1420
Waste Industries	Greenville NC	1-800-682-2492
Waste Industries Inc.	Durham NC	1-800-342-7920
Waste Industries Inc.	Oxford NC	1-800-682-2034
Waste Industries West Inc.	Graham NC	1-800-443-9952
Waste Leasing and Haulers	Syracuse NY	1-800-451-5765
Waste Management Michigan	Wayne MI	1-800-238-9278
Waste Management Of Baton Rouge	Walker LA	1-800 654-0489
Waste Management of Charleston	Summerville SC	1-800-854-6199
Waste Management Of Iowa	Des Moines IA	1-800 232-1859
Waste Management of Leon County Inc.	Tallahassee FL	1-800-262-4398
Waste Management Of Midwest	Grand Rapids MI	1-800 538-3750
Waste Management Of Midwest	Lansing MI	1-800 968-1570
Waste Management of New Hampshire	Londonderry NH	1-800-443-5515
Waste Management of Ohio	Columbus OH	1-800-334-8930
Waste Management of Pennsylvania	Elizabethtown PA	1-800-634-4595
Waste Management of Scranton	Dunmore PA	1-800-222-2028
Waste Management of South Carolina	Spartanburg SC	1-800-525-3109
Waste Management of South Louisiana	Raceland LA	1-800-548-8597
Waste Management of South Louisiana		1-800-624-4136
Waste Management Of Southeast Texas	Victoria TX	1-800 729-3666
Waste Management Of Southeast Texas	Wharton TX	1-800 366-9795
Waste Management of Tennessee Valley	Ooltewah TN	1-800-346-5145
Waste Management Of the Bayous	Franklin LA	1-800 349-6991
Waste Processing Equipment	Rainsville AL	1-800-225-6458
Waste Reduction Inc.	New Castle IN	1-800-452-5940
Waste Reduction Systems Inc.	West Jordan UT	1-800-858-9330
Waste to Energy Inc.	Graceville FL	1-800-637-3530
Waste Water Systems	Lilburn GA	1-800-828-9045
Water & Sewage Supply	Columbia MO	1-800-722-2330
Wellington Ltd.	Lindenwold NJ	1-800-242-1785
Western Waste Industry	Corona CA	1-800 858-8886
Whiskaway	Ft Wayne IN	1-800-255-3597
Whiskaway Inc	Ft Wayne IN	1-800 635-1103
Whitney And Whitney	Reno NV	1-800 368-1865
Wilson Refuse	Montrose CO	1-800-325-1797
Winter Welding and Machine	York PA	1-800-472-4250
Wood Robbie Haz Mat Transport	Dolomite AL	1-800-356-7457

WASTE PAPER & RECYCLING

Cook Paper Recycling Corp	Kansas City MO	1-800-654-2354
D C Intercell	Baltimore MD	1-800-875-0406
Southeastern Fibers Inc	Tucker GA	1-800-833-2472

WATER CONSERVATION

Applied Polymers	Rancho Cucamonga CA	1-800-232-9202
Con-Tech Industries	Carlsbad CA	1-800-446-5765
Cycle H2O	Camp Verde AZ	1-800-292-5342
Enviro-Check	Orlando FL	1-800-845-5036
Fluid Conservation Systems	Austin TX	1-800-531-5465
Hartman Enterprises	Palo Alto CA	1-800-421-7246
International Environmental Solutions Inc.	Clearwater FL	1-800-972-8348
Mellette County Conservation District	White River SD	1-800-235-0803
Trademark Sales and Manufacturing	Neenah WI	1-800-622-1818
Transtech Inc.	Incline Village NV	1-800-323-6111
US Water News Inc.	Halstead KS	1-800-251-0046

WATER FILTRATION AND PURIFICATION

Accufilter Inc	Blue Springs MO	1-800-336-1577
Agwa Systems	Burbank CA	1-800-473-9426
Air & Water International	Cannon Falls MN	1-800-538-1005
American Fil-Ter-Co Inc.	Kansas City MO	1-800-833-3586
American Products and Services Inc.	Marshfield MA	1-800-872-2770
American Pure Spring Water Sales	Orange NJ	1-800-423-9420
American Water Products	Fountain Valley CA	1-800-255-7124
American Water Purification Inc	Wichita KS	1-800-669-4758
Anabiosis Technologies	Yelm WA	1-800-233-4136
Applied Industrial Services	Cumming GA	1-800-443-5411
Aqua City Inc	Mountain View CA	1-800-962-5516
Aqua Media	Seattle WA	1-800-441-7634
Aqua Sun Inc	Minden NV	1-800-334-0082
Aqua Systems Inc.	Indianapolis IN	1-800-447-5582
Aqua Technology Water Stores	San Jose CA	1-800-478-7342
Aquavitae Costa	Mesa CA	1-800-468-4823
AR Systems	Myrtle Beach SC	1-800-382-3442
B & B Chlorination Inc.	Albert City IA	1-800-331-4808
Bauer Soft Water	South Bend IN	1-800-922-1211
Bayo's Water Vending Co	Swoyersville PA	1-800-237-2296
Best of Everything	Columbia SC	1-800-228-2063
Best Water H/B Associates	Falls Church VA	1-800-562-2994
Best Water Purification Systems	Cherry Hill NJ	1-800-441-0503
Butler Water Conditioning	Fairfield OH	1-800-232-1488
Carbon Sales	Wilkes-Barre PA	1-800-233-8355
Carico	Ft Lauderdale FL	1-800-422-7426
Cellini Purifications Systems Inc.	Westfield MA	1-800-628-7528
Chambers Pure Water Systems	Arrington TN	1-800-543-1464
Charlton Well	Charlton MA	1-800-338-6665
Chem-Serv	Thousand Oaks CA	1-800-468-8806
Chlor-Serv Inc.	San Dimas CA	1-800-338-5149
Clean Filters Co. Inc.	Lipan TX	1-800-453-4510
Clear Improvement Associates	Springfield MO	1-800-532-8272

Cranco Distributors/Everpure Water Filters	Edenton NC	1-800-428-7844
Crystal Clear	Westbrookville NY	1-800-433-9553
Crystal King	Middleville MI	1-800-243-5464
Culligan Funk Water Quality	Eagleville PA	1-800-535-6047
Culligan International Company	Northbrook IL	1-800-285-5442
Culligan Water Conditioning	Oklahoma City OK	1-800-299-0108
Culligan Water Conditioning	Miami FL	1-800-527-8150
Culligan Water Conditioning	Middletown OH	1-800-762-4119
Cuzn Water Filtration Systems Inc.	Fayetteville AR	1-800-345-7873
Dayton Water Systems	Dayton OH	1-800-424-9250
Deans Water Purification	Greenup KY	1-800-222-4097
Dew Con Enterprises	Grand Rapids MI	1-800-922-9426
Dunn Brothers Water Center	Shippenville PA	1-800-356-3866
Eastern Water Conditioning	Wilmington NC	1-800-962-3470
Eastlab Corp	New Bern NC	1-800-634-3134
Econeco Inc	Richland MI	1-800-472-9426
Ecowater Systems	Woodbury MN	1-800-545-1780
EES Corp	Sugar Land TX	1-800-621-9189
Engineering Specialties	Bensenville IL	1-800-854-4982
Environmental Products USA Inc.	Englewood FL	1-800-828-2447
Envirosafe Water Filters	Dalllas TX	1-800-628-9521
Equinox Independent Distributors	Sarasota FL	1-800-317-7873
Ever Pure Inc.	Downers Grove IL	1-800-552-6552
Everpure Inc		1-800-323-7873
Filter Cartridge & Media Water Purifica		1-800-463-2101
Filtration Marketing Service	Lutherville MD	1-800-628-3128
Flodur Entrps Distributor For Multi-Pure	Winston-Salem NC	1-800-577-1459
Flowright	Issaquah WA	1-800-932-6859
Fluid Systems	San Diego CA	1-800-525-4369
General Ecology Inc.	Lionville PA	1-800-441-8166
Global Health Products	Grapevine TX	1-800-227-2008
Harris Water Conditioning	Grabill IN	1-800-264-2195
Harris Water Conditioning	Grabill IN	1-800-264-2205
Harris Water Conditioning	Grabill IN	1-800-264-2206
Health & Water Inc.	Virginia Beach VA	1-800-523-6388
Hydro Life	Elkhart IN	1-800-626-7130
Hydro Quess Inc	Concord CA	1-800-468-8420
Hydro-Health Corp	Lexington MA	1-800-572-2151
Hydrotechnology	Valencia CA	1-800-356-1836
Indian Springs Water Co of Nevada	Sparks NV	1-800-345-2140
Integrity Filtration	Ipswitch MA	1-800-382-9930
Introdel Inc	Itasca IL	1-800-323-4772
Ionpure Tech For Western States	Tempe AZ	1-800-235-7739
Ionpure Technology	New Britain CT	1-800-322-6577
IW Technologies	San Diego CA	1-800-321-4989
J R C Industries	Hillside IL	1-800-979-1900
Joseph Pagani Independent Equinox Repr.	Chicago IL	1-800-566-2820
Kelco Water Engineering Inc.	Yuma AZ	1-800-365-3526
Ken Walsted & Associates	Lakeland FL	1-800-582-4483
Kinetico Inc/Engineered Systems Division	Newbury OH	1-800-633-5530
Kinetico Waterplant	Newbury OH	1-800-432-1166
Lake Whitney Water Co. Inc.	Whitnet TX	1-800-852-0418
Lapure Water Products	Madiera Beach FL	1-800-787-5300
Levy Limehouse Belinger Hill Water Inc.	Beaufort SC	1-800-842-9744
Lifestream Water Systems	Long Beach CA	1-800-468-5426
Magg-Flow	Sandy OR	1-800-221-9066
Mearlin Ionization Systems		1-800-628-4328
Metal Enterprises	Pismo Beach CA	1-800-282-1885

Minnehana Spring Water	Cleveland OH	1-800-367-0570
Mountain Filtration Systems	Clarksburg WV	1-800-552-3015
Mr. Water	Hastings NE	1-800-826-0610
Multi Pure Corp	Chatsworth CA	1-800-622-9206
Multi-Pure Drinking Water Systems	Van Nuys CA	1-800-644-2482
Multi-Pure Water Distributor	San Diego CA	1-800-533-7873
National Fluid Separators	St Louis MO	1-800-334-2460
New England Water System	Nashua NH	1-800-642-6397
Nimbus Water Systems	San Marcos CA	1-800-451-9343
Nimbus Water Systems	San Marcos CA	1-800-824-4915
North American Aqua Inc.	Vamdalia MI	1-800-833-5553
Northwest Backflow	Silverton OR	1-800-962-5879
NSA	Overland Park KS	1-800-474-1011
Oceanus	Ocala FL	1-800-447-0750
Optimum Waters	Billings MT	1-800-523-3212
Oritex Corp	Alhambra CA	1-800-831-2923
Oxygenics	Morgantown NC	1-800-680-0098
Ozone Research & Equip	Glendale AZ	1-800-446-0823
Pacific Rim Water	Sumner WA	1-800-525-7444
Peck Water Systems Inc.	N Canton OH	1-800-841-5994
Pep Process Efficiency Products	Chatsworth CA	1-800-243-4583
Polanetics	Southhampton PA	1-800-426-6013
Polymetrics	Weymouth MA	1-800-237-0778
Polymetrics	Weymouth MA	1-800-447-9094
Polymetrics	Weymouth MA	1-800-448-9465
Polymetrics	San Diego CA	1-800-445-5630
Polymetrics	Colorado Springs CO	1-800-447-9094
Polymetrics	S Windsor CT	1-800-231-2613
Polymetrics	Sunnyvale CA	1-800-338-7659
Polymetrics	Weymouth MA	1-800-448-9465
Precious Resources	Stockton CA	1-800-833-7757
Pro Flo Products	Wayne NJ	1-800-325-1057
Pur/Recovery Engineering Inc	St Louis Park MN	1-800-845-7873
Pura	Provo UT	1-800-292-7872
Pure Water Inc.	Lincoln NE	1-800-842-5805
Pure Water Systems	Castro Valley CA	1-800-848-4420
Pure Water Unlimited	Wisconsin Rapids WI	1-800-535-5778
Pure Water Works Inc.	Traverse City MI	1-800-248-7873
Pureco	Round Rock TX	1-800-531-9410
Puremark USA Inc.	W Caldwell NJ	1-800-982-5985
Puro Corp of America	Maspeth NY	1-800-336-7876
Quality Flow	Wheeling IL	1-800-227-5432
Quality Health Systems/Multi-Pure	Boulder CO	1-800-947-4450
Rain Soft Water Services	Ft Pierce FL	1-800-233-8884
Rainsoft	Vienna WV	1-800-456-0075
Reasco	Arvada CO	1-800-722-1229
Reverse Osmosis Engineering	Midland TX	1-800-774-2837
Richard Wycoff	Milford MI	1-800-258-6981
Richbourg Enterprises	Pensacola FL	1-800-942-3009
RR Watts Enterprises	Bennetts Field SC	1-800-647-0867
Schooner's International	Waco TX	1-800-845-4423
Shaklee Bestwater	N Potomac MD	1-800-445-8551
Sierra Pure Waters	Grand Rapids MI	1-800-952-6349
Stranco Inc.	Bradley IL	1-800-882-6466
Sun River Innovations Limited	Lexington KY	1-800-331-8988
Sun River Innovations Limited	Lexington KY	1-800-331-8988
Sure Way Systems	Syracuse NY	1-800-622-1002
Tarn Pure USA	Burr Ridge IL	1-800-635-7873

Teal Intl	Buffalo MN	1-800-222-6614
The Water Store	Nazareth PA	1-800-225-5897
Total Control Inc.	W Milford NJ	1-800-292-0919
Tremco	Barbourville KY	1-800-354-7892
Tri-Co	Georgetown TX	1-800-622-0950
Triad Water Systems International Inc.	Tracy CA	1-800-458-6600
Unifilt	Wilkes-Barre PA	1-800-752-3899
United Standard Of Maryland	Belle Aire MD	1-800-253-2692
UV Water Purification	Orchard Park NY	1-800-342-9731
Water & Life Products	Atlanta GA	1-800-543-3426
Water 2000	Trenton GA	1-800-426-7873
Water and Filtration Systems	Andrews IL	1-800-847-4247
Water Authority	Boca Raton FL	1-800-426-6529
Water Conscious	Thousand Oaks CA	1-800-633-9283
Water Distillers Inc.	Amblers PA	1-800-368-2162
Water Marque Inc.	Rockville MD	1-800-426-8818
Water Point Systems	Sunrise FL	1-800-226-9958
Water Spout	Dallas TX	1-800-847-7873
Water Systems Development/Pure-Tek	Mt Clemens MI	1-800-748-0434
Water Systems for Healthier Lives	N Miami FL	1-800-345-0765
Water Technologies Corp	Plymouth MN	1-800-637-1244
Watercoolers Etc.	Flushing NY	1-800-242-9384
Waterline Tech	Mansfield OH	1-800-522-3949
Waterwise Inc.	Leesburg FL	1-800-874-9028
Wettec	W Palm Beach FL	1-800-654-3893
YTT International	Hagerstown MD	1-800-847-2118
Zyzatech Water Systems	Seattle WA	1-800-633-3080

WATER PURIFIERS

General Ecology - First Need	Exton PA	1-800-441-8166
Katadyn USA Inc.	Scottsdale AZ	1-800-950-0808
PUR Water Purifiers		1-800-548-0406

WATER TESTING - SEE ALSO LABORATORIES

American Environmental Laboratories Inc.		1-800-522-0094
American Home Water Test Service	Syosset NY	1-800-852-3838
BioVir Lab Inc.	Benecia CA	1-800-442-7342
Chemetrics	Calverton VA	1-800-356-3072
Coast Medical Laboratory	Coos Bay OR	1-800-892-3195
Continental Hydrodyne Systems	S Lebanon OH	1-800-543-9283
Enviro Systems Control	Orange Park FL	1-800-451-8740
H2OK Inc.	Tallahassee FL	1-800-257-2738
HF Scientific	Ft Myers FL	1-800-798-2116
Hydro Group	Cochranton PA	1-800-331-5651
Hydro-Analysis Associates	Kutztown PA	1-800-622-6424
Industrial Cleaning Service	Girard OH	1-800-545-5531
Johnson Well Equip	Pensacola FL	1-800-874-8685
Leadetectors	Tahoe CA	1-800-453-2368
Mellish Waterblasting Equip Inc	Deerfield Beach FL	1-800-828-0408
Orbeco Analytical Systems	Farmingdale NY	1-800-922-5242
Ozark Analytical Water Laboratory & Serv.	Sulphur Springs AR	1-800-835-8908
Prosser Lab	Effort PA	1-800-213-6880
Sonford Samplers	Minneapolis MN	1-800-992-7267
Southeast Water Production	Blairsville GA	1-800-553-1264

Southern Well & Recovery	Owings MD	1-800-782-1536
Sprite Industries	Fullerton CA	1-800-327-9137
Suburban Water Testing Labs	Temple PA	1-800-433-6595
Taylor Technologies	Sparks MD	1-800-638-4776
Texas Tapwater Testing	Houston TX	1-800-241-7264
Truespin Auger Mktg	Corona CA	1-800-642-8437
US Water Testing	Kempton PA	1-800-837-8426
Water Testing Through Sears	St Paul MN	1-800-426-9345
Watertest Corp	Manchester NH	1-800-426-8376
Wells Waters and Gasses	Wise VA	1-800-435-8492

WATER TREATMENT

A Multi-Pure Drinking	Riverside CA	1-800-874-7873
Aeqrx Technologies Ltd	Warwick RI	1-800-331-2758
Aeration Industries	Chaska MN	1-800-328-8287
Aeration Industries International Inc.	Chaska MN	1-800-543-4475
Air-O-Lator Corp	Kansas City MO	1-800-821-3177
Alliance Group Inc.	Milwaukee WI	1-800-648-7339
Alloy Hardfacing & Engineering	Minneapolis MN	1-800-328-8404
American Alpha Water Conditioning	Dellroy OH	1-800-422-4449
American Development	Vanceboro NC	1-800-842-0764
American Sigma	Medina NY	1-800-635-4567
American Water Science	Cleveland OH	1-800-362-8086
American Water Treatment	Las Vegas NV	1-800-852-5005
Americleer	Ocala FL	1-800-237-8597
Amtek	Sheboygan WI	1-800-222-7558
Antec Inc.	Mt Pleasant SC	1-800-821-6698
Applied Biochemists	Milwaukee WI	1-800-558-5106
Applied Biochemists	Milwaukee WI	1-800-558-5106
Aqua Dynamics Corp Scalewatcher	Pompano Beach FL	1-800-241-7899
Aqua Dynamics Group	Adamsville TN	1-800-331-3338
Aqua Dynamics Group	Adamsville TN	1-800-882-0608
Aqua Labs	Amesbury MA	1-800-343-0213
Aqua Labs	Amesbury MA	1-800-892-0819
Aqua Magnetics International Inc.		1-800-328-2843
Aqua Resources	Garden Grove CA	1-800-732-9330
Aqua Treatment Services Inc	Harrisburg PA	1-800-338-2323
Aqua-Treet Associates Inc.	Spencer NY	1-800-582-6286
Aquakleer	Mountain View CA	1-800-962-5516
Aquarion Management Services	Monroe CT	1-800-832-3794
Aquatec	Merrilville IN	1-800-342-7646
Aquatec Chemical International	N Branch NJ	1-800-828-2331
Aquatech Of Florida Inc.	St Petersburg FL	1-800-343-1227
Aquatek Water Conditioning	Churubusco IN	1-800-552-4682
Arpco Pump Service	Houston MO	1-800-398-2583
Arts John	Scottsdale AZ	1-800-582-4179
Atlanco Of Ohio	Cortland OH	1-800-626-5842
Atlantic Filter Of Polk County	Lakeland FL	1-800-642-6548
Atlantic Water Products	Mays Landing NJ	1-800-228-3781
Axchem	Manistee MI	1-800-632-9823
Aylor Aqua Service Inc.	Dyersburg TN	1-800-527-4550
BA Ward & Associates	Houston TX	1-800-635-9420
Benz Industrial	Midlothian VA	1-800-235-6137
Benzsay and Harrison	Delanson NY	1-800-247-3798
Beta Technology Inc	Cambridge MD	1-800-638-9566
Beverage Suppiers Inc.	Thomasville NC	1-800-334-1877

Beyar Metropolitan Water District	San Antonio TX	1-800-552-5172
Biogenesis	Hope AR	1-800-327-1484
Bioprime Limited	Norwich VT	1-800-366-9141
Blankenship And Associates Inc.	Oviedo FL	1-800-432-5892
Bonco Manufacturing	Jefferson GA	1-800-442-6626
Bowman Instrument and Filter Service Co.	Bowman GA	1-800-628-1185
Browne Lab	Chattanooga TN	1-800-421-2436
By George	Avon OH	1-800-874-6367
Calgon Vestal	St Louis MO	1-800-648-9005
Capital Control	Colmar PA	1-800-523-2553
Capital Controls	Phoenix AZ	1-800-356-1381
Capital Controls Co Inc	Colmar PA	1-800-527-1998
Capital Controls Co. Inc.	Colmar PA	1-800-344-7968
Capital Controls Co. Inc.		1-800-523-2595
Carbon Sales	Wilkes-Barre PA	1-800-233-8355
Cassedy	Escondido CA	1-800-562-6304
Cassidy Water Conditioning	Lowell MA	1-800-428-8001
Chardon Labs	Columbus OH	1-800-282-3412
Chardon Labs	Columbus OH	1-800-848-9526
Charger Water Conditioning Inc.	Mokena IL	1-800-642-4274
Charger Water Treatment Products	Morton Grove IL	1-800-642-4274
Chemco	Warwick RI	1-800-221-5628
Chemicator Products	San Clemente CA	1-800-421-4822
Chemico International	Corpus Christi TX	1-800-272-4997
Chlorinators	Palm City FL	1-800-327-9761
Cincinnati Technical Services Inc.	Cincinnati OH	1-800-331-8467
Cl2 Equipment	Lancaster TX	1-800-336-1126
Clean Water Systems	Columbia Falls MT	1-800-543-0671
Coastline	Plano TX	1-800-443-9779
Columbiana County Water & Sewer Dept	Lisbon OH	1-800-843-1752
Continental Technologies Inc.	Little River KS	1-800-847-5874
Continental Water Systems	Gainesville FL	1-800-342-1103
Crystal Clear Distillers	Westbrookville NY	1-800-433-9553
Culligan Dutchess Putnam	Poughkeepsie NY	1-800-334-0220
Culligan Industrial Water	Glenwood IL	1-800-241-3224
Culligan of Angleton	Angleton TX	1-800-245-4675
Culligan Water Conditioning	Logansport IN	1-800-682-7638
Culligan Water Conditioning	Coloma MI	1-800-442-2802
Culligan Water Conditioning	Bethel CT	1-800-221-5522
Culligan Water Conditioning	Delphi IN	1-800-854-5002
Culligan Water Conditioning	Kingsland GA	1-800-441-7061
Culligan Water Conditioning	Reynoldsburg OH	1-800-541-6000
Culligan Water Conditioning of Lehigh	Allentown PA	1-800-258-4202
Culligan Water Conditioning of Wayne Co	Sodus NY	1-800-448-4418
Dacar Chemical	Pittsburgh PA	1-800-223-8875
Davis Waterworks Equipment	Bullhead City AZ	1-800-732-1318
Dayton Water Systems	Dayton OH	1-800-524-6019
Dayton Water Systems	Cincinnati OH	1-800-535-5585
De Wallace Technical Sales Inc.	Worcester MA	1-800-838-0969
Delta Water Lab	Lubbock TX	1-800-858-4586
Dennis Chlorination Service	Baltimore MD	1-800-443-1755
Diamond Water Systems	Holyoke MA	1-800-245-6601
Dias	Kalamazoo MI	1-800-332-3427
DMP Corp	Rock Hill SC	1-800-845-5019
Eagelbrook	Schererville IN	1-800-428-3311
Ecolochem	Norfolk VA	1-800-446-8004
Ecowater Systems	York Springs PA	1-800-542-8649
Electrocatalytic Inc.	Union NJ	1-800-553-5228

Electronic Water Treaters	Effort PA	1-800-344-5939
EMTEC	Lake Forest CA	1-800-438-3683
Environ	Newport Beach CA	1-800-523-2093
Environmental Pre Treatment Systems	Mt Holly NC	1-800-428-3774
Environmental Systems Trading Co	Menlo Park CA	1-800-327-4399
Environmental Training Consultants Inc.	Corvallis OR	1-800-451-7747
Envirotec Operating Services Inc.	Birmingham AL	1-800-331-8482
Envirozone Industries	Fort Pierce FL	1-800-438-4081
Equipment Enterprises	Atlanta GA	1-800-221-3681
ETL Water Central	Lexington MO	1-800-428-1385
EWT Dayton Water Systems	Louisville KY	1-800-880-8841
Faw John Sales Inc.	Vinton VA	1-800-942-9094
Feed Rite Controls	Fond Du Lac WI	1-800-231-0358
Fini Enterprises	Celina TX	1-800-441-2659
Fluid Tech International	Ft Wayne IN	1-800-348-1835
Fogle Pump & Supply	Colville WA	1-800-533-6518
Frank J Anfosso & Associates, Inc.	Sugar Land TX	1-800-982-1696
Friar Chemical Co	Lavalette WV	1-800-535-2155
Gene Mktg	Trumbull CT	1-800-426-4556
GS Service Corp	Montpelier IN	1-800-225-3087
H2O Purification	Fairfield CT	1-800-426-4556
Hadley Industries	Ludington MI	1-800-345-4227
Hague Water Conditioning Inc.	Jackson NJ	1-800-822-0192
Handi Products Inc	Grand Rapids MI	1-800-635-9645
Hart Bill Associates Inc.	Jackson MI	1-800-292-5922
Hawkins Water Tech	Middlebury IN	1-800-360-9292
HE Anderson Co Inc.	Muskogee OK	1-800-331-9620
Heffernan Soft Water	Hillsdale MI	1-800-426-0210
Howco Environmental Service	St Petersburg FL	1-800-334-6926
Hycor	Lake Bluff IL	1-800-624-8415
Hydrodynamics	Bogalusa LA	1-800-327-9304
Hydromatix Inc.	Brea CA	1-800-221-5152
Hydrotec Services Inc.	Altoona PA	1-800-822-3080
IMI World Marketing	Oak Lawn IL	1-800-544-6884
Imperial West Chemical	Antioch CA	1-800-321-4922
Impulse Regeneration Services	San Antonio TX	1-800-292-1011
Inland Aqua Tech	Spokane WA	1-800-331-3314
Ionics	Watertown MA	1-800-446-6427
Island Park Water Co	Macks Inn ID	1-800-547-7338
Itek	Hollywood FL	1-800-972-7885
Jacobs Well Water Treatment & Service	Cutler OH	1-800-997-9355
JBI Water and Waste Water Equipment	Burlingame CA	1-800-524-9378
John Faw Sales Inc	Vinton VA	1-800-942-9094
Judson May	Gardena CA	1-800-675-7525
Kem Mfg	Cincinnati OH	1-800-424-8790
Kemtune Inc.	Ft Wayne IN	1-800-348-0999
Keystone Lab	Decatur AL	1-800-526-7767
Kiss Intl	Las Vegas NV	1-800-551-5477
Kjell Water Consultants	Janesville WI	1-800-356-0422
Krystal Kleer	Scottsdale AZ	1-800-423-6889
Lafever Water Conditioning	Cuba NY	1-800-968-3622
Lakewood Instruments	Phoenix AZ	1-800-228-0839
Lakos Separators	Fresno CA	1-800-344-7205
Lakos Separators	Fresno CA	1-800-742-1850
Leco Water Treatment	Rural Hall NC	1-800-238-7540
Lindsay/Eco Water Systems	Fremont NE	1-800-642-8043
Matt's Softwater	Celina OH	1-800-788-6288
Matt-Son Inc.	Barrington IL	1-800-833-5593

McNeil Co	Tallahassee FL	1-800-342-7174
Microbe Masters	Elkton MD	1-800-847-2847
Mid America Supply	St Louis MO	1-800-443-3796
Monarch Water Systems	Xenia OH	1-800-553-3697
Monte Terme Water Systems Inc	Fremont CA	1-800-793-1014
Moody Bro's Inc.	Houston TX	1-800-533-3048
Mr Water	Wilkes-Barre PA	1-800-928-3794
MSI	Birmingham AL	1-800-343-1121
Multi-Pure Drinking Water Systems	Chatsworth CA	1-800-622-9205
Murdock G A Inc	Madison SD	1-800-568-4301
Natures Choice Water	El Campo TX	1-800-962-8551
NSA	Solano Beach CA	1-800-428-8777
NTU Technologies	Davis CA	1-800-342-6733
Ontrac Environmental Inc.	Richmond CA	1-800-992-9113
Osby Water Conditioning	Hebron IN	1-800-552-6729
Osmonics	Minnetonka MN	1-800-351-9008
Ozone Pure Water Inc.	Sarasota FL	1-800-633-8469
P E R Enterprises	Alpine CA	1-800-659-8818
Pella Water Conditioning	Pella IA	1-800-272-6212
Pennsylvania American Water Co Central	Milton PA	1-800-222-2487
Piedmont Quality Water	Madison GA	1-800-221-0835
Polanetics Inc.	Southhampton PA	1-800-426-6013
Pollution Technology Systems	Garland TX	1-800-662-4010
Polybac Corp	Topeka KS	1-800-358-9382
Polymetrics	Colorado Springs CO	1-800-822-7659
Polymetrics Inc.	Richmond VA	1-800-637-1308
Preferred Environmental Products Inc.	Plymouth MI	1-800-544-0389
Production Sales	Atlanta GA	1-800-652-6311
Professional Water Systems	Richfield CT	1-800-432-6897
ProH2O Inc.	Roanoke VA	1-800-831-8224
Puritan Water Conditioning Inc	Crawfordsville IN	1-800-367-6340
Puritron Inc.	Hialeah FL	1-800-554-5707
Quality Flow	Northbrook IL	1-800-227-5432
Rainborn Inc	Dalton GA	1-800-360-5537
Rainshow'r/Pacific Environmental	San Gabriel CA	1-800-243-8775
Rainsoft of Des Moines	Des Moines IA	1-800-652-9511
RBC Services Inc.	Milwaukee WI	1-800-232-7011
Reynolds Water Conditioning	Detroit MI	1-800-572-9575
RG Systems Inc.	Tampa FL	1-800-423-7808
RGF Environmental	Princeton IL	1-800-992-5743
RGF Inc.	W Palm Beach FL	1-800-633-7743
Rochester-Midland Corp.	Franklin MA	1-800-677-7757
Roco Corp	Brentwood MD	1-800-445-1127
Scale Watcher	Oxford PA	1-800-416-3183
Schreiber	Trussville AL	1-800-438-7335
Sea Recovery	Gardena CA	1-800-354-2000
Serv A Care	Butler PA	1-800-922-1421
Servisoft	Winchester VA	1-800-468-7400
Servisoft of Middlefield	Burton OH	1-800-424-7638
Ses	Rockford IL	1-800-468-4319
Shubert Associates	Clear Lake IA	1-800-642-4162
Sidener Environmental Services	St Louis MO	1-800-528-2887
Sierra Environmental Solutions Company	Hickory NC	1-800-775-4363
Smith & Loveless	Lenexa KS	1-800-922-9048
Smith Ecological Systems	Rockford IL	1-800-468-4319
South Sound Culligan	Tacoma WA	1-800-826-5984
Southeastern Water Treatment Co	Pineville WV	1-800-243-7681
Southern Water Treatment	Greenville SC	1-800-873-1755

Southwest Service and Equipment	Owasso OK	1-800-462-0950
Specialty Filtration Products	Perkasie PA	1-800-843-2018
Stafford Culligan Enterprises	Jasper AL	1-800-221-9314
Standard Water Systems	Ottawa Lake MI	1-800-452-9858
Stoner Quality Water Inc.	Roanoke VA	1-800-438-5595
Suma Water Consultants	Muncy MI	1-800-331-7763
Surge Water Conditioning	Hopkins MN	1-800-634-5307
Swan Water System	FLINT MI	1-800-292-7926
Sweet Water Labs	Winfield IL	1-800-450-2929
Sweetwater Technologies	Laguna Hills CA	1-800-426-2428
The Taulman Co	Atlanta GA	1-800-241-0179
Thornton Group	Albion PA	1-800-392-2368
TMG Service Inc.	Maple Valley WA	1-800-562-2310
Tri Mountain Water Conditioning	Ortonville MI	1-800-978-3800
Tri-Tron International	S Lyon MI	1-800-851-7860
Triangle Chemical Co	Chapel Hill NC	1-800-643-6485
Tru-Water	Casselberry FL	1-800-831-9283
Ultra Pure Systems Water Purification	SMinneapolis MN	1-800-634-1455
Ultra-Flo Systems Inc	Inglewood CA	1-800-348-5872
Unifilt Corp	Zelienople PA	1-800-223-2882
Unocal Chemicals	Fullerton CA	1-800-323-8647
US Filter/NW	Redmond WA	1-800-982-7135
Village Marine TEC	Gardena CA	1-800-421-4503
Wallace and Tiernan	Belleville NJ	1-800-628-0897
Wallace and Tiernan	Belleville NJ	1-800-822-1230
Walling Chemical	Sioux Falls SD	1-800-843-2326
Waltron	Murray Hill NJ	1-800-242-7353
Wascon Inc.	Livingston TN	1-800-952-4236
Waste Water Systems Inc.	Corona CA	1-800-628-3396
Water Authority Inc The	Oldsmar FL	1-800-426-6529
Water Conditioning	Knoxville TN	1-800-524-4441
Water Doctor	Selbyville DE	1-800-528-2250
Water Dynamics Inc	Gresham OR	1-800-635-1576
Water Point Systems	Ft Worth TX	1-800-582-5994
Water Quality Services	Carson City NV	1-800-748-6859
Water Rights Erie And Venus PA	Venus PA	1-800-262-0192
Water Rite	Venus PA	1-800-426-7483
Water Supplies	Ashland OH	1-800-822-9355
Water Tamer	Roseburg OR	1-800-762-3341
Water Tech Inc.	Twin Falls ID	1-800-367-3250
Water Treatment and Control	Pensacola FL	1-800-826-7699
Water Treatment Equipment	Yarmouth ME	1-800-328-7328
Whittaker Water Management Systems	Riverside CA	1-800-334-2161
Winston Chemicals Inc.	Bixby OK	1-800-331-9099
Wolcott Water Systems	Columbia MO	1-800-325-0104

WATER TREATMENT EQUIPMENT & SUPPLIES

AAA Molybdenum Products	Broomfield CO	1-800-443-6812
Accucast	Bellmead TX	1-800-433-6885
Air Gap Intl Inc	Irvine CA	1-800-433-5754
Aqua-Aid Systems	Keene NH	1-800-252-8484
B & R Industries	Mesa AZ	1-800-525-1302
Beacon Water Equip	Chenango Bridge NY	1-800-648-5248
Carolina Water Conditioners	Greensboro NC	1-800-332-5189
Charger Water Treatment Products Inc	Venice FL	1-800-690-5009
Clearwater Tech Inc	San Luis Obispo CA	1-800-262-0203

Commercial Water Systems	Houston TX	1-800-628-4785
Culligan Water Conditioning Corp	Fairmont MN	1-800-722-0598
Culligan Water Conditioning Of Orange Cty	Sun Valley CA	1-800-458-9572
Culligan Water Conditioning	Ocala FL	1-800-233-2063
Culligan Water Conditioning	Hagerstown MD	1-800-451-7512
Culligan Water Conditioning	E Hartford CT	1-800-842-1116
Culligan	Northbrook IL	1-800-428-2827
Dakota Scientific	Des Moines IA	1-800-240-3330
Davis Meter & Supply	Orlando FL	1-800-432-8513
Drillers Service Inc	Smithfield NC	1-800-662-5143
East Kentucky Miracle Water	Pikeville KY	1-800-521-7867
Gordon Brothers	Salem OH	1-800-331-7611
Graver Chemical	Glasgow DE	1-800-533-6623
Hydro Bionix	Grass Valley CA	1-800-274-1440
Idreco USA Ltd	N Oxford MA	1-800-552-1594
Inn Exchange Products Inc	Chicago IL	1-800-621-7113
J & J Water Cleaning	Modena NY	1-800-606-6611
J Karp & Sons Inc	Factoryville PA	1-800-344-0587
La Framboise Well Driving & Water Service	Thompson CT	1-800-624-2327
Layne Central	Pensacola FL	1-800-356-3824
Performance Filters Of North America	Cincinnati OH	1-800-548-5785
Phillippe Water Equip	Anderson IN	1-800-232-2962
Pressure Vessel Technologies	Santa Fe Springs CA	1-800-826-5748
Pumps & Controls	Muskogee OK	1-800-331-7338
Purolite Co Inc-Fax, The	Bala-Cynwyd PA	1-800-260-1065
Rain Soft Of Franklin County	Shady Grove PA	1-800-323-7321
Rainsoft-Distributed By Southern Ohio	Wilmington OH	1-800-258-8261
Related Products	Dayton OH	1-800-735-2833
Scale Watcher	Oxford PA	1-800-337-2253
South Jersey Water Conditioning Service	Bridgeton NJ	1-800-321-7592
Springsoft	Belvidere IL	1-800-358-3879
T D I Processing Inc	Pompano Beach FL	1-800-952-5933
TPT Inc	Pelham AL	1-800-848-1563
Techtronics	Ashland OH	1-800-521-9210
Town & Country Water Treatment Systems	Erie PA	1-800-833-9959
Water Inc	El Segundo CA	1-800-322-9283
Water Master Inc	Harrisburg PA	1-800-572-9135
Water Plus	Birmingham AL	1-800-842-9979
Water Soft	Ashland OH	1-800-462-3790
Water Soft	Ashland OH	1-800-732-8749
Watermaker Corp	Garland TX	1-800-420-7873
Wolverine Water Treatment	Kincheloe MI	1-800-521-3040

WATER WELL DRILLING & EQUIPMENT

Alpine Drilling	Omak WA	1-800-400-5143
Bishop Well And Pump Service Inc	Tifton GA	1-800-342-9345
Brown's Whsle Pump Supply	Gettysburg PA	1-800-722-7930
Chrisley Water Well Service	Mexia TX	1-800-533-7224
Country Town Drilling	Carlton WA	1-800-410-9355
Double J Water Well Drilling	Groesbeck TX	1-800-545-3386
Ed Birkmeier Well Drilling Ltd	New Lothrop MI	1-800-638-5104
Far West Pump	Willcox AZ	1-800-545-4261
Flexcon Industries	Randolph MA	1-800-527-0030
Geo Hydro Data	Tehachapi CA	1-800-351-0507
Golden Gate Well Drill & Water Condition.	Naples FL	1-800-768-9355
Griffin Dewatering	Omaha NE	1-800-999-3038

Houston Well Screen Co	Park City UT	1-800-732-0196
Howard Smith Screen Co Inc	Houston TX	1-800-527-4772
Layne Inc-Safety Division	Kansas City KS	1-800-272-7978
M W Cornell & Sons Inc Well Drilling	Williamston MI	1-800-551-6489
Morris Industries	Pompton Plains NJ	1-800-835-0777
Northwest Well Drilling	Salem OR	1-800-669-6404
Olson Brothers Well	Eau Claire WI	1-800-257-0324
Petersons Well Drilling County Line	Irons MI	1-800-882-4963
Sperry Drilling	Berlin PA	1-800-296-3487
Trilogy Controls Inc	Mountain View CA	1-800-243-6956
Victor Pipe & Steel	Winfield MO	1-800-264-6315
Welenco Inc	Bakersfield CA	1-800-445-9914

WEATHER FORECASTING

1-800-Weather	Atlanta GA	1-800-932-8437
Accu Weather Inc.	State College PA	1-800-881-3349
Accu-Weather Inc	State College PA	1-800-438-9847
Air Routing International-"Canada"	Houston TX	1-800-238-3126
Jeppese Data Plan	Los Gatos CA	1-800-358-6468
KTVI TV	St Louis MO	1-800-544-5884
Kavouras Inc	Minneapolis MN	1-800-328-2278
Meteorological Evaluation Services	Amityville NY	1-800-952-2052
Micro Forecasts Inc	Hood River OR	1-800-955-1645
Official NOAA Weather Radio Forecasts	Nashville TN	1-800-662-6622
Savannah County Airport Automated Weather	Savannah TN	1-800-972-6080
Strategic Weather Services	Wayne PA	1-800-882-5881
Wilkens Weather Technologies	Houston TX	1-800-503-5811
ZFX Weather Information By Fax	Westborough MA	1-800-876-1232

Environmental Job Titles & Positions

Senior, Intermediate, entry-level, technician, paid intern and volunteer intern positions may exist for every title or position. Most professional positions require BS/MS in a related field or discipline. Technician positions may be degreed or non-degreed. Professional, management and senior technical positions, typically require relevant experience and demonstrated leadership and management skills.

ENVIRONMENTAL ENGINEERING

Air Quality Engineers
Civil Engineers
Chemical Engineers
Design Engineers
Environmental Engineers
Hazardous Waste Engineers
Hazardous Waste Design Engineers
Industrial Engineers
Industrial Process Engineers
Mechanical Engineers
Sanitary Engineers
Water Treatment Engineers

EARTH SCIENCE & ENGINEERING

Cartographers
Geochemists
Geographers
Geologists
Geomorphologists
Geophysicists
Geophysical Engineers
Geotechnical Engineers
Ground Water Modelers
Ground Water Scientists
Ground Water Specialists
Hydrodynamics
Hydrogeologists
Hydrologists

ENVIRONMENTAL SCIENCES

Environmental Analysts
Environmental Auditors/Inspectors
Environmental Compliance Specialists
Environmental Impact Analysts
NEPA Specialists
Environmental Scientists
Environmental Specialists
Environmental Protection Specialists
Air, Water, Ground Water, Soils Scientists
Air, Water, Ground Water, Soils Specialists
Laboratory Analysts & Technicians
NEPA Studies/Integration Specialists
Regulatory Specialists
Permit Writers
Permitting Specialists
RCRA Closure Specialists
Remedial Investigation Specialists
UST Specialists

ENVIRONMENTAL RESTORATION

Construction Superintendents Construction Managers
Construction QA Specialists
Decontamination Specialists
Decommissioning Specialists
Drilling Services/Support Specialists
Facility Surveillance Specialists
Facility Maintenance Specialists
Field Investigation Specialists

Operable Unit Managers
Pipe/Pipeline Specialists
Plant Operators
Plant and Facility Managers
Plant and Facility Technicians
Remediation Engineers
Remediation Specialists
Tank Specialists
Underground Storage Tank Specialists

HAZARDOUS MATERIALS

Asbestos Removal Specialists
Toxic Chemical Specialists
Toxic Substance Specialists
Toxic Material Specialists
Hazardous Materials Managers
Hazardous Materials Specialists
Hazardous Materials Spill & Emergency Specialists

HAZARDOUS WASTE

Waste Designation Specialists
Waste Acceptance Specialists
Environmental Compliance Coordinators
Hazardous Waste Managers
Hazardous Waste Management Specialists
Hazardous Waste Specialists
Hazardous Waste Compliance Specialists
Permit Writers
Pollution Prevention Specialists
TSD Facility Managers
TSD Facility Operators
TSD Facility Maintenance Specialists
Technical Operations Managers
Technical Operations Specialists
Waste Management Specialists
Waste Minimization Specialists

INDUSTRIAL HYGIENE/SAFETY

Emergency Preparedness Specialists
Occupational Safety Specialists
Industrial Compliance Specialists
Industrial Hygienists
Industrial Safety Specialists
Health Physics Technologists
Fire Protection Engineers
Nuclear Safety Specialists
Radiological Safety Specialists

LABORATORY SERVICES

Analytical Chemistry Scientists & Specialists
Chemists
Organic Chemists
Radiochemists
Mixed Waste Radiochemists
GC/HPLC/GC/MS Scientists & Specialists
Sampling Technologists
Sample Tracking Specialists
Sample Preparation Specialists

Field and Mobile Laboratory Specialists
Data Validation Specialists
Data Management Specialists
Laboratory QA/QC Specialists
Analytical Protocol Specialists
Analytical QA/QC Specialists

LIFE SCIENCES

Air Quality Scientists
Aquatic Ecologists
Archeologists
Cultural Resource Specialists
Biologists
Chemists
Dispersion Modelers
Ecologists
Environmental Impact Analysts/Specialists
Environmental Scientists
Fishery Biologists
Human Health Risk Specialists
Life Scientists
Meteorologists
NEPA Specialists
Program Managers & Coordinators
Resource Planners
Risk Assessment Specialists
Social Scientists
Terrestrial Ecologists
Toxic Modelers
Toxicologists
Water Quality
Water Resources Specialists
Wetlands Scientists/Specialists

PROJECT AND CONSTRUCTION MANAGEMENT

CAD Department Managers
CAD Specialists
Project Managers/Specialists
Project Control Specialists
Construction Inspectors
Construction Managers
Cost and Schedule Engineers
Estimators
Construction Specialists

DESIGN SUPPORT

Corrosion Engineers
Electrical Engineers
Instrumentation Engineers
Mechanical Engineers
Metallurgical Engineers
Nuclear Engineers
Quality Assurance Engineers & Specialists
Quality Control Engineers and Specialists
Radiological Engineers
Structural Engineers
Value Engineers

SOLID WASTE

Landfill Design Specialists
Landfill Closure Specialists
Incinerator Design Specialists
Pollution Prevention Specialists
Recycling Specialists
Solid Waste Facility Managers
Solid Waste Facility Operators
Waste Minimization Specialists

SPECIALTY TECHNICIANS

CAD/CADD Specialists
Drafters
Engineering Technicians
Technical Writers
Specialty Writers
Technical Editors
Operations and Maintenance Technicians

SUPPORT DISCIPLINES

Architects
Planners
ADP Specialists
Economists
Labor Relations
Legal
Paralegal
Training Specialists
Security Specialists

ADMINISTRATIVE

Administrative Managers
Administrative Specialists
Controllers
Data Base Managers
Data Base Specialists
Filing
Graphics Designers
Information Systems Management Specialists
Information Systems Specialists
Library Specialists
Mail Services Administrators
Reproduction Specialists
Printing Specialists
Telecommunications Specialists

PERFORMANCE ASSESSMENT

Compliance Inspectors
Audit and Appraisal Specialists
ES & H Auditors, Inspectors & Surveillance Specialists
Financial Audit Specialists

PROJECT CONTROL

Project Control Managers
Project Control Specialists
Scheduling Specialists
Cost Control Specialists
Cost Engineers
Cost Estimators

PROGRAM MANAGEMENT & SUPPORT

Program Integration Specialists
Program Interface Specialists
Configuration Management Specialists
Systems Engineers & Specialists
Systems Analysts
Planning Managers
Financial Planners
Strategic Planners
Tracking Managers

QUALITY ASSURANCE, QUALITY CONTROL & QUALITY MANAGEMENT

Quality Engineers
Self-assessment Specialists

Quality Planners/Coordinators
Contract and Subcontract QA/QC Specialists
TQM Specialists
Process Improvement Coaches

TECHNOLOGY DEVELOPMENT

Technology Development Managers/
Technology Development Specialists
Technology Development Engineers
Technology Development Scientists
Technology Transfer Specialists
Technology Demonstration Specialists
Environmental Educators
Outreach Specialists
Training Specialists

BUSINESS

Accountants
Accounts Payable
Budget Analysts
Financial Analysts
Contract & Subcontract Administrators
Business Development Managers & Specialists
Technical Sales Representatives
Marketing Managers/Specialists
Federal Marketing Specialists
Contract Specialists
Payroll Specialists

Purchasing Specialists
Procurement Specialists
Property Management Specialists
Real Estate Specialists
Small Business Specialists, Liaisons & Coordinators
Small Disadvantaged Business Specialists &
Coordinators
Minority Business Development Specialists

COMMUNICATIONS/PUBLIC RELATIONS

Communications Specialists
Community Relations Specialists
Community Involvement Specialists
Internal Communications Specialists
Journalists
Media Relations Specialists
Photographers

HUMAN RELATIONS - PERSONNEL HUMAN RESOURCES

Compensation Specialists
Employee Relations
Equal Opportunity Specialists
Human Resources Managers
Human Resources Specialists
Labor Relations Specialists
Training Specialists/Coordinators
Recruiting Specialists/Coordinators

Environmentally Related Federal Government Positio
Classifications by Occupational Group and Series

GS-000 - MISCELLANEOUS OCCUPATIONS GROUP - NOT CLASSIFIED ELSEWHERE

Safety and Occupational Health Management	GS-018
Safety Technician	GS-019
Community Planning	GS-020
Community Planning Technician	GS-021
Outdoor Recreation Planning	GS-023
Park Ranger	GS-025
Environmental Protection Specialist	GS-028
Environmental Protection Assistant	GS-029
Security Administration	GS-080
Fire Protection and Prevention	GS-081
U. S. Marshall	GS-082
Police	GS-083
Nuclear Materials Courier	GS-084
Security Guard	GS-085
Security Clerical and Assistance	GS-086
Guide	GS-090
Foreign Law Specialist	GS-095
General Student Trainee	GS-099

GS-100 - SOCIAL SCIENCE, PSYCHOLOGY AND WELFARE GROUP

Social Science	GS-101
Social Science Aid and Technician	GS-102
Economist	GS-110
Economics Assistant	GS-119
Geography	GS-150
Civil Rights Analysis	GS-160
History	GS-170
Psychology	GS-180
Sociology	GS-184
Recreation Specialist	GS-188
Recreation Aid and Assistant	GS-189
General Anthropology	GS-190
Archeology	GS-193
Social Science Student Trainee	GS-199

GS-200 - PERSONNEL MANAGEMENT AND INDUSTRIAL RELATIONS GROUP

Personnel Management	GS-201
Military Personnel Management	GS-205
Personnel Staffing	GS-212
Position Classification	GS-221
Occupational Analysis	GS-222
Salary and Wage Administration	GS-223
Employee Relations	GS-230
Labor Relations	GS-233
Employee Development	GS-235
Mediation	GS-241
Apprenticeship and Training	GS-243
Labor Management Relations Examining	GS-244
Contractor Industrial Relations	GS-246
Wage and Hour Compliance	GS-249
Equal Employment Opportunity	GS-260
Federal Retirement Benefits	GS-270
Personnel Management Student Trainee	GS-299

GS-300 GENERAL ADMINISTRATION, CLERICAL AND OFFICE SERVICES GROUP

Miscellaneous Administration and Program	GS-301
Work Unit Supervising	GS-313
Computer Operating	GS-332
Computer Specialist	GS-334
Program Management	GS-340
Administrative Officer	GS-341
Support Services Administration	GS-342
Management and Program Analysis	GS-343
Logistics Management	GS-346
Equal Opportunity Compliance	GS-360
Equal Opportunity Assistance	GS-361
Telecommunications Processing	GS-390
General Communications	GS-392
Administration and Office Support Student Trainee	GS-399

GS-400 - BIOLOGICAL SCIENCES GROUP

General Biological Sciences	GS-401
Microbiology	GS-403
Biological Science Technician	GS-404
Pharmacology	GS-405
Agricultural Extension	GS-406
Ecology	GS-408
Zoology	GS-410
Physiology	GS-413
Entomology	GS-414
Toxicology	GS-415
Plant Protection Technician	GS-421
Botany	GS-430
Plant Pathology	GS-434
Plant Physiology	GS-435
Plant Protection and Quarantine	GS-436
Horticulture	GS-437
Genetics	GS-440
Range Conservation	GS-454
Range Technician	GS-455
Soil Conservation	GS-457
Soil Conservation Technician	GS-458
Irrigation System Operation	GS-459
Forestry	GS-460
Forestry Technician	GS-462
Soil Science	GS-470
Agronomy	GS-471
Agricultural Management	GS-475
General Fish and Wildlife Administration	GS-480
Fishery Biology	GS-482
Wildlife Refuge Management	GS-485
Wildlife Biology	GS-486
Animal Science	GS-487
Home Economics	GS-493
Biological Science Student Trainee	GS-499

GS-500 ACCOUNTING AND BUDGET GROUP

Financial Administration and Program	GS-501
Financial Management	GS-505

Accounting	GS-510
Auditing	GS-511
Budget Analysis	GS-560
Financial Management Student Trainee	GS-599

GS-600 - MEDICAL, HOSPITAL AND PUBLIC HEALTH GROUP

General/Health Science	GS-601
Public Health Program Specialist	GS-685
Sanitarian	GS-688
Industrial Hygiene	GS-690
Consumer Safety	GS-696
Environmental Health Technician	GS-698
Medical and Health Student Trainee	GS-699

GS-700 - VETERINARY MEDICAL SCIENCE GROUP

Veterinary Medical Science	GS-701
Animal Health Technician	GS-704
Veterinary Student Trainee	GS-799

GS-800 - ENGINEERING AND ARCHITECTURE GROUP

General Engineering	GS-801
Engineering Technician	GS-802
Safety Engineering	GS-803
Fire Protection Engineering	GS-804
Materials Engineering	GS-806
Landscape-Architecture	GS-807
Architecture	GS-808
Construction Control	GS-809
Civil Engineering	GS-810
Surveying Technician	GS-817
Engineering Drafting	GS-818
Environmental Engineering	GS-819
Construction Analyst	GS-828
Mechanical Engineering	GS-830
Nuclear Engineering	GS-840
Electrical Engineering	GS-850
Computer Engineering	GS-854
Electronics Engineering	GS-855
Electronics Technician	GS-856
Biomedical Engineering	GS-858
Aerospace Engineering	GS-861
Naval Architecture	GS-871
Ship Surveying	GS-873
Mining Engineering	GS-880
Petroleum Engineering	GS-881
Agricultural Engineering	GS-890
Ceramic Engineering	GS-892
Chemical Engineering	GS-893
Welding Engineering	GS-894
Industrial Engineering Technician	GS-895
Industrial Engineering	GS-896
Engineering and Architecture Student Trainee	GS-899

GS-900 - LEGAL AND KINDRED GROUP

Law Clerk	GS-904
General Attorney	GS-905
Paralegal Specialist	GS-950
Contract Representative	GS-962
Land Law Examining	GS-965
Legal Occupations Student Trainee	GS-999

GS-1000 - INFORMATION AND ARTS GROUP

General Arts and Information	GS-1001
Exhibits Specialist	GS-1010
Museum Curator	GS-1015
Museum Specialist and Technician	GS-1016
Illustrating	GS-1020

Office Drafting	GS-1021
Public Affairs	GS-1035
Language Specialist	GS-1040
Photography	GS-1060
Audiovisual Production	GS-1071
Writing and Editing	GS-1082
Technical Writing and Editing	GS-1083
Visual Information	GS-1084
Editorial Assistance	GS-1087
Information and Arts Student Trainee	GS-1099

GS-1100 - BUSINESS AND INDUSTRY GROUP

General Business and Industry	GS-1101
Contracting Series	GS-1102
Industrial Property Management	GS-1103
Property Disposal	GS-1104
Purchasing	GS-1105
Procurement Clerical and Technician	GS-1106
Property Disposal Clerical and Technician	GS-1107
Public Utilities Specialist	GS-1130
Trade Specialist	GS-1140
Agricultural Program Specialist	GS-1145
Agricultural Marketing	GS-1146
Industrial Specialist	GS-1150
Production Control	GS-1152
Financial Analysis	GS-1160
Crop Insurance Administration	GS-1161
Insurance Examining	GS-1162
Loan Specialist	GS-1163
Realty	GS-1170
Appraising and Assessing	GS-1171
Building Management	GS-1176
Business and Industrial Student Trainee	GS-1199

GS-1200 - COPYRIGHT, PATENT AND TRADE-MARK GROUP

Patent Technician	GS-1202
Copyright	GS-1210
Copyright Technician	GS-1211
Patent Administration	GS-1220
Patent Advisor	GS-1221
Patent Attorney	GS-1222
Patent Classifying	GS-1223
Design Patent Examining	GS-1224
Copyright and Patent Student Trainee	GS-1299

GS-1300 - PHYSICAL SCIENCES GROUP

General Physical Science	GS-1301
Health Physics	GS-1306
Physics	GS-1310
Physical Science Technician	GS-1311
Geophysics	GS-1313
Hydrology	GS-1315
Hydrology Technician	GS-1316
Chemistry	GS-1320
Metallurgy	GS-1321
Astronomy and Space Science	GS-1330
Meteorology	GS-1340
Meteorological Technician	GS-1341
Geology	GS-1350
Oceanography	GS-1360
Navigational Information	GS-1361
Cartography	GS-1370
Cartographic Technicians	GS-1371
Geodesy	GS-1372
Land Surveying	GS-1373
Geodetic Technician	GS-1374
Forest Products Technology	GS-1380
Food Technology	GS-1382

Textile Technology	GS-1384
Photographic Technology	GS-1386
Document Analysis	GS-1397
Physical Science Student Trainee	GS-1399

GS-1400 - LIBRARY AND ARCHIVES GROUP

Librarian	GS-1410
Librarian Technician	GS-1411
Technical Information Services	GS-1412
Archivist	GS-1420
Archives Technician	GS-1421
Library and Archives Student Trainee	GS-1499

GS-1500 - MATHEMATICS AND STATISTICS GROUP

Actuary	GS-1510
Operations Research	GS-1515
Mathematics	GS-1520
Mathematics Technician	GS-1521
Mathematical Statstician	GS-1529
Statistician	GS-1530
Statistical Assistant	GS-1531
Cryptography	GS-1540
Cryptanalysis	GS-1541
Computer Science	GS-1550
Mathematics and Statistics Student Trainee	GS-1599

GS-1600 - EQUIPMENT, FACILITIES AND SERVICES GROUP

General Facilities and Equipment	GS-1601
Facility Management	GS-1640
Laundry and Dry Cleaning Plant	GS-1658
Equipment Specialist	GS-1670
Equipment and Facilities Management Trainee	GS-1699

GS-1700 - EDUCATION GROUP

General Education and Training	GS-1701
Education and Training Technician	GS-1702
Educationa and Vocational Training	GS-1710
Training Instruction	GS-1712
Vocational Rehabilitation	GS-1715
Education Program	GS-1720
Public Health Educator	GS-1725
Education Research	GS-1730
Education Services	GS-1740
Instructional Systems	GS-1750
Education Student Trainee	GS-1799

GS-1800 - INVESTIGATION GROUP

General Inspection, Investigation and Compliance	GS-1801
Compliance Inspection and Support	GS-1802
General Investigating	GS-1810
Criminal Investigating	GS-1811
Game Law Enforcement	GS-1812
Air Safety Investigating	GS-1815
Mine Safety and Health Investigating	GS-1822
Aviation Safety	GS-1825
Agricultural Commodity Warehouse Examining	GS-1850
Consumer Safety Inspection	GS-1862
Food Inspection	GS-1863
Public Health Quarantine Inspection	GS-1864
Import Specialist	GS-1889
Investigation Student Trainee	GS-1899

GS-1900 - QUALITY ASSURANCE, INSPECTION AND GRADING GROUP

Quality Assurance	GS-1910
Agricultural Commodity Grading	GS-1980
Agricultural Commodity Aid	GS-1981
Quality Inspection Student Trainee	GS-1999

GS-2000 - SUPPLY GROUP

General Supply	GS-2001
Supply Program Management	GS-2003
Supply Clerical and Technician	GS-2005
Inventory Management	GS-2010
Distribution Facilities and Storage Management	GS-2030
Packaging	GS-2032
Supply Student Trainee	GS-2099

GS-2100 - TRANSPORTATION GROUP

Transportation Specialist	GS-2101
Transportation Clerk and Assistant	GS-2102
Transportation Industry Analysis	GS-2110
Railroad Safety	GS-2121
Motor Carrier Safety	GS-2123
Highway Safety	GS-2125
Traffic Management	GS-2130
Travel Series	GS-2132
Transportation Loss and Damage Claims Examining	GS-2135
Transportation Operations	GS-2150
Air Traffic Control	GS-2152
Marine Cargo	GS-2161
Aircraft Operations	GS-2181
Air Navigation	GS-2183
Aircrew Technician	GS-2185
Transportation Student Trainee	GS-2199

Yellow Pages Listing of Regulated Environmental Industry Categories

Agricultural Consultants
Air Pollution Control
Architects
Asbestos Consultants and Testing
Asbestos Removal and Abatement Services
Associations
Building Inspection Services
Bulletin and Directory Boards
Business Consultants
Chemical Disposal Service
Chemicals - Retail
Chemicals - Storage and Handling
Chemicals - Wholesale & Manufacturers
Consultants:
    Business Consultants
    Communications Consultants
    Economic Research and Consulting Services
    Education Consultants
    Environmental and Ecological Services
    Fishery Consultants
    Forestry Consultants
    Industrial Hygiene Consultants
    Management Consultants
    Real Estate Consultants
    Safety Consultants
    Waste Management Consultants
Drilling and Boring Contractors
Economic Research and Consulting Services
Education Consultants
Environmental and Ecological Services
Electric Power and Light Companies
Employment Agencies
Energy Management and Conservation Services
Engineers:
    Agricultural
    Chemical
    Civil
    Construction
    Consulting
    Corrosion
    Environmental
    Geological
    Geotechnical
    Hazardous Waste
    Industrial Waste
    Marine
    Mechanical
    Nuclear
    Structural
    Water Supply
Environmental, Conservation and Ecological Organizations
Environmental Consultants
Environmental and Ecological Services

Environmental Impact Statements
Executive Search Consultants
Fishery Consultants
Government Offices:
    City, Village and Township
    County
    State
    United States
Health Agencies
Industrial Hygiene
Laboratories:
    Analytical
    Research and Development
    Testing
Land Planners
Landfills-Sanitary
Legal Research
Management Consultants
Mediation Services
Metals
Mining
Nuclear Fuels
Nuclear Services
Oils - Wastes
Oils - Recycling Centers
Paralegals
Pest Control Services
Planning Consultants
Real Estate Consultants
Recycling Centers
Research Consultants
Research and Development Service
Researchers - Independent
Resource Consultants
Safety Consultants
Scientists - Consulting
Soil Testing
Solar Energy
Solvents
State Government
US Government
Waste Disposal - Chemical and Radioactive
Waste Reduction, Disposal and Recycling Service - Industrial
Water Pollution Control
Water Purification Services
Water Supply Systems
Water Well Drilling and Service
Water Works Equipment & Services
Weed Control Service

INDEX

About the Author

Paul Krupin is the publisher of Direct Contact Publishing in Kennewick, Washington.

He is presently an Environmental Program Manager with the U. S. Department of Energy, at the Hanford Site, in Richland, Washington. He has over 20 years of diverse professional experience in environmental protection, pollution control, and natural resource management at complex industrial, nuclear and waste management facilities, including employment with the State of Oregon, Department of Environmental Quality Water Quality Division, the U.S. Department of Interior, Bureau of Land Management, Idaho, the U.S. Department of Agriculture, Forest Service, Oregon and Idaho, several law firms, and private industry. His first environmental job was as a seismic geophysical exploration technician (or doodlebugger, as it was then known) in Colorado.

Born in New York City, he was raised in Franklin Square, on Long Island. He has received a Bachelors Degree in Environmental Biology from the University of Colorado in Boulder, Colorado, a Masters Degree in Geography from Oregon State University in Corvallis, Oregon, and a Juris Doctor, with a special certificate in Dispute Resolution, from Willamette University, in Salem, Oregon.

Environmental sciences have long fascinated him, and early in his career he got hooked on getting others into the field. During the summer of 1993, Mr. Krupin's hobby of helping people learn how to get environmental jobs with government and industry led to an invitation to be a guest speaker at the Environmental Careers Organization Ninth Annual Conference in Tampa, Florida. He noticed that most job seekers and many working professionals did not know how to contact prospective employers and colleagues in the field. So he came up with the idea of a directory of toll-free numbers for the environmental industry as a solution to the costs and challenges faced by the hundreds of enthusiastic and dedicated college students, career changers and professionals who attended the conference. Mr. Krupin says he derives great personal satisfaction from helping people achieve their goal of working and being successful in this field. He believes that to these people it is more than just a job. It's a value. It provides meaning to their lives.

Mr. Krupin has also published The National 8(a) Minority & Disadvantaged Directory, FED-Pro: The Federal Procurement Database, and DOD-Pro: The Department of Defense Procurement Database, the Toll-Free Hunting and Archery Directory, and a Toll-Free Fishermen's Directory. He is 43 years old. He is a once-upon-a-time avid fisherman, who loves the Pacific Northwest. His wife Nancy is a Registered Dietitian. They have two highly energetic children, and two very laid back cats.